OXFORD MEDICAL PUBLICATIONS

Delirium in the Elderly

Delirium in the Elderly

James Lindesay
Senior Lecturer in Psychogeriatrics

Alastair Macdonald
Senior Lecturer in Psychogeriatrics

and

Ian Starke
Senior Lecturer in Geriatric Medicine

United Medical and Dental Schools
Guy's Hospital
London, UK

Oxford   New York   Tokyo
OXFORD UNIVERSITY PRESS
1990

Oxford University Press, Walton Street, Oxford OX2 6DP
Oxford   New York   Toronto
Delhi   Bombay   Calcutta   Madras   Karachi
Petaling Jaya   Singapore   Hong Kong   Tokyo
Nairobi   Dar es Salaam   Cape Town
Melbourne   Auckland
and associated companies in
Berlin   Ibadan

Oxford is a trade mark of Oxford University Press

Published in the United States
by Oxford University Press, New York

British Library Cataloguing in Publication Data
Lindesay, James
Delirium in the elderly.
1. Old persons. Mental disorders
I. Title   II. Macdonald, Alastair 1947–   III. Starke, Ian
618.97689

ISBN 0–19–261862–8

Library of Congress Cataloging in Publication Data
Lindesay, James.
Delirium in the elderly / James Lindesay, Alastair Macdonald, Ian Starke
p. cm. – (Oxford medical publications)
Includes bibliographical references.
Includes index.
1. Delirium in the elderly.   I. Macdonald, Alastair (Alastair John
Douglas) II. Starke, Ian.   III. Title.   IV. Series.
[DNLM: 1. Cognition Disorders – in old age.   2. Delirium – in old
age. WM 204 L745d]
RC520.7.L55   1990   618.97'689 – dc20   90-7656

ISBN 0–19–261862–8

Set by CentraCet, Cambridge
Printed in Great Britain by
Courier International Ltd., Tiptree, Essex

Preface

Despite the considerable and growing interest in the disorders of old age in recent years, delirium in the elderly remains relatively neglected. In part, this is because it is not yet considered to be particularly important; after all, it is transient, usually 'quiet', and powerful sedatives are available to control any disturbed behaviour that might occur. In fact, the transience and quietness of much delirium in the elderly mean that it is often missed, or dismissed as dementia, and important opportunities for early identification and correction of the underlying causes are lost. As for management, sedation is still very much a two-edged sword in delirium, compounding difficulties more often than resolving them, and the various psychological and environmental alternatives are rarely practicable on a busy ward. There is as yet no specific treatment.

Another reason for the current neglect of delirium in the elderly is its borderline nature: it is not quite medicine, not quite psychiatry. It is interesting how often problems of diagnosis and management of delirium manifest themselves as battles between physicians and psychiatrists over 'difficult' patients. These border disputes have unfortunately deflected attention and interest away from the real issue, which is our lack of knowledge about all aspects of delirium in old age.

Two of the authors of this book are psychiatrists, and one is a geriatrician, so we hope to have avoided the narrowness of vision that might have resulted from approaching this topic from one direction only. Our aim has been not only to gather together the useful knowledge about delirium in the elderly, but also to convey the extent of our ignorance, and to indicate how it might be overcome. The sooner this edition is out of date, the better.

London   J.E.B.L.
July 1990   A.J.D.M.
                                                                     I.D.S.

Contents

1 The concept of delirium

Physicians have put a difference betwixt frenzy and madness, imagining madness to be only an infection and perturbation of the foremost cell of the head, whereby imagination is hurt; but frenzy to extend further, even to offend reason and the memory, and is never without fever. Galen calls it an inflammation of the brains, or films thereof, mixed with a sharp fever.

Adams (1861)

Introduction

The term 'delirium' chosen for use throughout this book was one of several possible candidates. There are plenty to choose from: 'confusion', 'toxic confusion', 'acute organic brain syndrome', 'acute confusional state', 'acute brain failure', 'acute organic psychosis', 'acute organic reaction', 'metabolic encephalopathy' – even 'pharmacotoxic psychosis' (Danielczyk 1984). Lishman (1987) has recounted the different ways terms such as these have been used, opting to subsume most of them under the heading of 'acute organic reaction'. However, he hints that the term 'organic' may also have its own peculiar difficulties. Grimley Evans (1982) has described three different subgroups of 'acute confusional state' (functional psycho-syndromes, decompensated dementia and delirium) that blur the organic/ functional divide. Examinees for the Membership of the Royal College of Psychiatrists in the UK are informed very definitely by Davison (1989) that three syndromes (delirium, dysmnesic syndrome and 'quasifunctional syndromes') comprise 'acute organic brain syndromes'. Neurologists, on the other hand, currently favour a distinction between the 'acute confu-sional state' and 'acute agitated delirium' (e.g. Mori and Yamadori 1987).

As Lipowski (1983) has commented, there is considerable terminological chaos surrounding attempts to characterize those states where cognitive function is transiently disrupted by a major physical or personal insult. Our choice of the term 'delirium' has been determined by the following considerations. First, despite certain shifts in its use, the notion of 'delirium' has been one of the most stable in the history of psychiatry. Consider this discussion of 'the Deliriums attending acute or other Distempers' by Hartley (1749):

In these a disagreeable State is introduced into the nervous System by the bodily Disorder, which checks the Rise of pleasant Associations, and gives Force and Quickness to disgustful ones . . . a vivid Train of visible Images forces itself upon the Patient's Eye; and that from a Disorder in the Nerves and Blood-vessels of the Eye itself, or from one in the Brain, or one in the alimentary Duct, or, which is most probable, from a Concurrence of all these . . . That delirious Persons have such Trains forced upon the Eye from internal Causes, appears probable from hence, that when they first begin to be delirious, and talk wildly, it is generally at such times only as they are in the Dark, so as to have all visible Objects excluded; for, upon bringing a Candle to them, and presenting common Objects, they recover themselves, and talk rationally, till the Candle be removed again.

This account of delirium with visual hallucinations is perfectly comprehensible to the modern clinician.

The second reason for favouring 'delirium' is that terms such as 'confusion' have been absorbed much more into everyday lay language, and unlike similarly vulgarized terms such as 'depression' or 'schizophrenia' they have not attracted a redeemingly specific technical use. Lipowski (1983) has argued strongly against the use of 'confusion' on these grounds. Since almost any cognitive impairment (and much besides) in the elderly is described by many professionals and informal carers as confusion, it is vital to separate transient, remediable states as much as possible from the aura of permanence that surrounds this word.

Berrios (1981) has provided a comprehensive review of the origins of the concepts of delirium and confusion. He draws attention to the difficulties that the seventeenth century Cartesian dualism of mind and body introduced into the previously uncomplicated understanding of the relationship between physical and mental disease. He also argues that the flux of ideas surrounding these concepts provide a better paradigm of nineteenth century psychiatric thought than paralysis of the insane. The word 'delirium' stems from the Latin roots de- (meaning off, away from) and lira (a ridge between ploughed furrows), indicating that the patient is 'off the ridge between furrows', that is, at some state parallel to but lower than normality. This metaphor emphasized the general view until the nineteenth century that delirium was primarily a disorder of the 'train' of thinking, a notion that persists today in expressions such as 'off the rails' and 'off his trolley'. Although in most psychiatric traditions the concept has widened to include affective and motor disturbances, the old idea of delirium as a specific disorder of logical thinking has persisted, notably in France. As Berrios has noted, the persistence of 'delirie' as a description of disordered thinking in French psychiatry has allowed the term 'confusion' to be applied exclusively to the behavioural, cognitive and emotional aspects of the acute organic states. For instance, 'bouffée delirante' in French psychiatric terminology shares only disordered thinking with the Anglo-American concept of delirium.

Formal psychiatric diagnostic systems

The most widely used systems in current use are the Ninth Edition of the International Classification of Diseases (ICD9) (World Health Organization (WHO) 1977) and the Third Edition of the Diagnostic and Statistical Manual of Mental Disorders (DSM-III) (American Psychiatric Association (APA) 1980). Both systems have recently been further revised: the DSM-III-R (APA 1987) has been published, and the Tenth Edition of the ICD (ICD10) is currently available in draft form. The ICD and DSM systems represent very different approaches to psychiatric classification.

The ICD system

The ICD9 codes are designed to allow clinicians from all over the world to categorize their patients with a minimum of formality or difficulty; there are few formal rules, and terms are not clearly defined. Delirium is coded under Transient Organic Psychiatric Conditions (293), a category that is mainly subdivided into 'Acute Confusional State' (293.1) and 'Subacute Confusional State' (293.2) (Table 1.1). These categories exclude delirium occurring in the context of senile dementia (290.3), and delirium tremens caused by alcohol withdrawal (290.1), but specifically include delirium in the course of cerebrovascular disease. This inclusion of cerebrovascular delirium may represent a persistance of old ideas concerning the aetiology of delirium in the elderly (see also Chapter 4).

Although ICD9 code 293 is to be used in cases where there is 'usually' some 'intra- or extracerebral toxic, infectious, metabolic or other systemic disturbance', and where the state is 'generally' reversible, it is still possible to use code 293 for cases where the syndrome of delirium is present, but has no ascertainable cause. Thanks to its utilitarian vagueness (for example, the state 'lasts hours or days'), ICD9 has partially escaped the criticism levelled against other definitions; in particular, that it is not possible to categorize patients until they are either dead or better.

However, this vagueness has one unfortunate result. The concept of 'senility', a hangover from a previous ICD category of 'symptoms, senility and ill-defined conditions', still appears as an interesting vestigial remnant under the definition of Transient Organic Psychotic Conditions, and betrays the origins of this form of classification. Up to, and including, the Ninth Edition, the ICD was clearly designed to *follow* ordinary clinical practice, and if it seemed woolly, or if terms were used loosely, it was so that ordinary clinicians from a wide range of backgrounds could use the system with no discomfort (or much thought). 'Senility' may have almost disappeared, but the term 'confusion' is still enshrined by the ICD9 in

Table 1.1   ICD9 Transient organic psychotic conditions (293)

States characterized by clouded consciousness, confusion, disorientation, illusions, and often vivid hallucinations. They are usually due to some intra- or extracerebral toxic, infectious, metabolic, or other systemic disturbance and are generally reversible. Depressive and paranoid symptoms may also be present but are not the main feature.

Excludes: confusional state or delirium superimposed on senile dementia (290.3)

> dementia caused by: alcohol (291)
> arteriosclerosis (290.4)
> senility (290.0)

293.0 Acute confusional state
Short-lived states, lasting hours or days, of the above type.

Acute: delirium
infective psychosis
organic reaction
post-traumatic organic psychosis
psycho-organic syndrome
psychosis associated with endocrine, metabolic or cerebrovascular disorder

Epileptic: confusional state
twilight state

293.1 Subacute confusional state
States of the above type in which the symptoms, usually less florid, last for several weeks or longer, during which they may show marked fluctuations in intensity.

Subacute: delirium
infective psychosis
organic reaction
post-traumatic organic psychosis
psycho-organic syndrome
psychosis associated with endocrine or metabolic disorder

293.2 Other

293.9 Unspecified

codes 293.1 and 293.2, even though this word is already used indiscriminately by many lay people (and not a few doctors) to describe demented as well as delirious elderly patients. 'He's confused', and 'He's got an acute confusional state' should mean very different things, but, sadly, the crucial distinction between dementia and delirium has been confounded by the use of the word 'confusion' in the WHO classification.

Despite the recommendation of the WHO in favour of the continued use of 'acute confusional state' (1972), defined as a syndrome 'characterised by features of delirium' (implying that delirium is a particular sub-set of symptoms, rather than a syndrome in its own right), the draft ICD10 which appeared in 1987 marked a shift of emphasis, both in the adoption of definite rules for the categorization of patients, and, in the case of delirium, the abandonment of the term 'confusion' in a coded category (although not in the definitions), and the adoption of a syndrome-orientated approach. The ICD10 category F05, 'Delirium, other than Alcoholic' can be used when there is some impairment in *each* one of the areas termed consciousness, cognition, psychomotor function, sleep–wake cycle and emotion, but cannot be used if the patient is free from disturbances in any one area (Table 1.2). Delirium is defined as an 'aetiologically non-specific organic cerebral syndrome', a phrase which some might find contradictory, and it is distinguished from presumed non-organic states in which 'confusional features may be present'. Despite this, the diagnosis is permitted in the absence of a known organic cause, but cannot be made if the condition lasts more than six months, so that patients cannot be coded unless they have recovered or died.

The DSM system

DSM-III is a rule-orientated approach to classification which represented, in 1980, a major departure from its predecessor DSM-II, and from the nomenclature of ICD8 and ICD9. DSM-III makes a distinction between Organic Brain *Syndromes* and Organic Mental *Disorders*: the former, which include Delirium, are allocated without reference to aetiology, the latter are diagnosed if an Organic Brain Syndrome is present, together with evidence for the presence of a specific aetiological organic factor. Unfortunately, this careful distinction between syndrome and disorder breaks down in the case of delirium, since the DSM-III diagnostic criteria for this particular Organic Mental Syndrome (Table 1.3) include a stipulation (criterion E) that there should be evidence of a 'specific organic factor'. If this evidence is available, surely the delirium should qualify as an Organic Mental Disorder? Using DSM-III, if the delirium is caused by the action or withdrawal of a substance such as alcohol, barbiturates, amphetamines or phencyclidine, it may be classified further as a Disorder, but this is not

Table 1.2   Draft ICD10 delirium, other than alcoholic (F05):
diagnostic guidelines

Symptoms, mild or severe, must be present *in each one* of the following areas:

(1) impairment of consciousness and attention
(on a continuum from clouding to coma; reduced ability to direct, focus, sustain, and shift attention)

(2) global disturbance of cognition
(perceptual distortions, illusions and hallucinations – most often visual; impairment of abstract thinking and comprehension, with or without transient delusions but typically with some degree of incoherence; impairment of immediate recall and of recent memory but with relatively intact remote memory; disorientation for time as well as, in more severe cases, for place and person)

(3) psychomotor disturbances
(hypo- or hyperactivity and unpredictable shifts from one to the other; increased reaction time; increased or decreased flow of speech; enhanced startle reaction)

(4) disturbance of the sleep–wake cycle
(insomnia or, in severe cases, total sleep loss or reversal of the sleep–wake cycle; daytime drowsiness; nocturnal worsening of symptoms; disturbing dreams or nightmares which may continue as hallucinations after wakening)

(5) emotional disturbances
(e.g. depression, anxiety or fear, irritability, euphoria, apathy or wondering perplexity)

The onset must be rapid, the course diurnally fluctuating, and the total duration of the condition under six months.

---

possible for deliria due to other physical causes, such as infection, electrolyte disturbance or hypoxia. This paradoxical situation appears to be the result of the acceptance by DSM-III of certain traditions with regard to the classification of organic mental disorders, one of which is that any definition of delirium must presume an organic aetiology.

The revised version of DSM-III, DSM-III-R, incorporates significant changes in the diagnostic criteria for the syndrome of delirium (Table 1.4). Notable in this revision of the DSM-III criteria are the disappearance of the term 'clouding', and the option to diagnose delirium even when no 'specific organic aetiological factor' has been determined. However, criterion E2 still requires such a factor to be 'presumed' if the symptoms are not caused by a non-organic mental disorder.

Table 1.3 DSM-III diagnostic criteria for delirium

A: Clouding of consciousness (reduced clarity of awareness of the environment), with reduced capacity to shift, focus, and sustain attention to environmental stimuli.

B: At least two of the following:

   (1) perceptual disturbance: misinterpretations, illusions, or hallucinations;

   (2) speech that is at times incoherent;

   (3) disturbance of the sleep–wakefulness cycle, with insomnia or daytime drowsiness;

   (4) increased or decreased psychomotor activity.

C: Disorientation and memory impairment (if testable).

D: Clinical features that develop over a short period of time (usually hours to days) and tend to fluctuate over the course of a day.

E: Evidence, from the history, physical examination, or laboratory tests, of a specific organic factor judged to be aetiologically related to the disturbance.

Nosological problems

Two main problems exist for the taxonomist when dealing with delirium. First, the central distinction between the organic and the functional which underpins Western psychiatric classification is aetiological to a degree which far outstrips our actual understanding of the aetiology of most psychiatric disorders. This is not the place to discuss the reasons for this, but one of the consequences is that a classification system such as DSM-III that insists that delirium can only be diagnosed when a physical cause or process can be demonstrated is left with a residue of clinically identical states (such as Lipowski's 'pseudodelirium') that cannot be easily categorized. On the other hand, a system such as DSM-III-R or ICD10, by allowing the diagnosis of delirium in the absence of an obvious physical cause or process, may lead to difficulties in distinguishing delirium from acute schizophrenia or dissociative states; in short, it dismantles the central partition between organic and functional psychiatric disorders.

The second problem has to do with course. Delirium is usually recognized as having an abrupt onset, fluctuating course, and terminating either in recovery, death or a different state, such as dementia. Whilst the former

Table 1.4   DSM-III-R diagnostic criteria for delirium

A: Reduced ability to maintain attention to external stimuli (e.g. questions must be repeated because attention wanders) and to appropriately shift attention to new external stimuli (e.g. perseverates answer to a previous question).

B: Disorganized thinking, as indicated by rambling, irrelevant, or incoherent speech.

C: At least two of the following:

   (i) reduced level of consciousness (e.g. difficulty keeping awake during examination);

   (ii) perceptual disturbances: misinterpretations, illusions, or hallucinations;

   (iii) disturbance of the sleep–wake cycle with insomnia or daytime sleepiness;

   (iv) increased or decreased psychomotor activity;

   (vi) disorientation to time, place, or person;

   (v) memory impairment (e.g. inability to learn new material, such as the names of several unrelated objects after five minutes, or to remember past events, such as history of current episode of illness.

D: Clinical features develop over a short period of time (usually hours to days) and tend to fluctuate over the course of a day.

E: Either (1) or (2):

   (1) evidence from the history, physical examination, or laboratory tests of a specific organic factor (or factors) judged to be aetiologically related to the disturbance;

   (2) in the absence of such evidence, an aetiologic organic factor can be presumed if the disturbance cannot be accounted for by any nonorganic mental disorder, e.g. Manic Episode accounting for agitation and sleep disturbance.

---

two criteria may be established while the patient is ill, and so may usefully inform the assessment process, the final outcome is inevitably a matter of conjecture at the time of initial assessment. A classification system that insists that the outcome be known may be of retrospective value, for instance in administrative returns, but is useless for guiding clinical practice or allocating patients in a clinical trial. On the other hand, ignoring outcome will lead to difficulty in distinguishing delirium from other conditions with a protracted course, a distinction that is of enormous practical value and one that has been made from early times (Garrison 1929).

The continuing uncertainty over the nature of delirium is, in no small part, due to the problems (by no means peculiar to delirium) that arise from difficulty in distinguishing between delirium as a 'disease' and as a 'syndrome'. A syndrome (from Greek συν, meaning together, and δραμεῖν, to run) is a set of symptoms and signs that occur together with sufficient frequency to make the set recognizable. Some syndromes (Horner's syndrome, for example) are closely related to a specific chain of pathological events, whilst others (like the Patterson–Kelly–Brown syndrome) are merely random aggregations. Syndrome identification is a statistical rather than an aetiological process – the selection of the sample of patients confronting any physician may equally well conceal associations found in the general population as produce associations not previously recognized. As a rule, the former leads the physician to dismiss the existence of the syndrome, the latter leads to publication.

Viewed aetiologically, the syndrome concept is highly questionable; the same syndrome in two patients may be caused by different pathological processes (or none at all, if it is a chance association), and have different aetiologies. The two patients may have different courses, and respond differently to treatment. So of what value is the recognition of syndromes? The principal advantage of the syndrome is the substantially atheoretical perspective that it embodies. The separation of observation – both of individual symptoms and of groups of symptoms and signs – from aetiological, pathological, therapeutic and prognostic speculation emphasizes the important distinction between evidence and hypothesis. Of course, no system of medical examination can be truly atheoretical – the simple, almost semi-conscious act of examining the conjunctivae is imbued with considerable theoretical and aetiological expectations – but without any attempt at a detached catalogue of the patient's complaints and signs, and the recognition of characteristic clusters, the modern physician would flounder in a slough speculation. What primitive understanding of psychiatric disorders we have we owe to the painstaking cataloguing of phenomena in clinical settings, as well as their epidemiological equivalents, rather than to the innumerable theoretical systems that have been erected over the millenia.

The concept of disease has attracted much more attention than that of the syndrome, particularly in connection with psychiatric practice (Kendell 1975; Kräupl-Taylor 1976). For the most part, discussion has revolved more about when disease is present at all, rather than which disease is present. The disease concept is essentially similar to that of an hypothesis. Diagnosis of a disease differs from the recognition of a syndrome in that the former is a hypothesis about the patient – what has happened to the patient's physiological processes (and perhaps what caused it), what will happen (prognosis being the most valued ancient medical art), what treatments may change the prognosis, and what might be found at post-mortem. By contrast, to identify a syndrome is merely to describe the current state of the patient.

Delirium does not always fall happily into this schema. On the one hand, it may be regarded as a syndrome – the characteristic collection of signs and symptoms described in Chapter 2. However, if course and outcome are included in the definition of delirium (as they are in ICD10), delirium then becomes a disease category – that is, despite the presence of a characteristic collection of phenomena, the syndrome of delirium cannot be recognized unless the outcome is clear. Thus we can distinguish between two processes – the identification of the *syndrome* of delirium, irrespective of the aetiology, course or outcome, and the *diagnosis* of delirium, which involves the making of a predictive hypothesis.

In their use of the term 'delirium', many of the investigations referred to in this book confound the central distinction between delirium as a syndrome and delirium as a diagnosis. In particular, two criteria are frequently cited as necessary for delirium – that there must be an organic aetiology, and that the course must be transient. Lipowski (1983), by recognizing that the syndrome of delirium may be present in the absence of obvious physical cause, has coined the unfortunate term 'pseudodelirium'. Clearly, this term is meaningless if one espouses a syndromic view of delirium – one cannot accept a 'pseudosyndrome'. Again, if one insists on a physical cause before delirium can be said to be present, then it follows that the investigation of the relationship of 'organic' and 'non-organic' factors in delirium cannot be undertaken without tautology. Similarly, investigation of the course of delirium is meaningless if course is part of its definition.

Nosologies are human constructs, and must be judged by their usefulness in the various settings where they are deployed. On a practical level, clinicians often face delirium before a comprehensive physical examination of the patient is possible, without a clear idea of the distinction between the 'organic' and the 'functional', and always without prescience, and for them the syndrome concept of delirium would seem to be the most useful starting point.

Research assessments of delirium

If the conceptual framework of current formal classification systems is somewhat opaque on the subject of delirium, then that apparent in many of the studies discussed below is frequently Stygian. No research assessment procedure specifically designed to assess delirium has yet been reported. The better studies have attempted some form of standardized assessment of cognitive function, and the Mini-Mental State Examination (MMSE) (Folstein *et al.* 1975) is the most frequently used instrument as a marker of delirium. This is an interviewer-administered rating scale in which the patient's memory, orientation in time and place, figure copying, language, attention, and concentration are briefly assessed. Folstein and his colleagues, apparently alarmed at 'extraordinarily high' sensitivity and specificity coefficients, undertook a study comparing the MMSE with independent psychiatric assessment in 97 patients aged from 20 to over 80 years admitted to general medical wards (Anthony *et al.* 1982). The clinical assessment for dementia and delirium were based on DSM-III definitions, but they ignored the DSM-III criteria for delirium that specified the course and organic aetiology (see page 7), thus defining delirium firmly as a syndrome. Essentially the only criterion which distinguished delirium from dementia in their clinical assessment was the presence of altered consciousness. Clinical ratings nominated 13 patients with 'dementia', nine with 'delirium', and one with both. High levels of sensitivity, based on a cut-point of the MMSE of 23/24, were achieved against independent clinical assessment – but dementia and delirium were amalgamated as a single category. Given that the clinical assessments were syndrome-based, their results were not surprising, and comparable with those reported for other short tests of memory and orientation used in the study of cognitive decline. Folstein and his colleagues did not report validity data for delirium separate from dementia. However, they reported test–retest reliability of the MMSE separately for dementia and delirium, and this showed what might be seen as an encouraging difference – an unspecified reliability coefficient of 0.90 for 12 'demented' patients, and 0.56 for seven 'delirious' patients retested after one day. They attributed the low rates of test–retest reliability in the case of delirium to fluctuation in the patients' condition.

A validation of the Pfeiffer Short Portable Mental Status Questionnaire (SPMSQ) (Pfeiffer 1975) as a screening test for dementia and delirium by Erkinjuntti *et al.* (1987) produced similar results. The SPMSQ was found to have good specificity and sensitivity as a dementia screen in elderly community and in-patient samples, but its performance as a delirium screen in the patients was not as good, principally because of the variable clinical picture.

These studies raise interesting issues for the methodology of the study of delirium. Instruments such as the MMSE and SPMSQ are very similar to a number of measures of cognitive function used in clinical and epidemiological studies, such as the Organic Brain Syndrome scale of the CARE schedule (Gurland *et al.* 1977) or the Cognitive Assessment Scale of the Clifton Assessment Procedures for the Elderly (CAPE) (Pattie and Gilleard 1979), and are unlikely to be capable on a single administration of distinguishing between the deficits of transient and permanent deficits, on two counts.

First, the phenomenological overlap between transient and permanent states seems to be in the very areas of cognitive function that these measures assess (especially orientation and memory), and no attention has been paid to the phenomenology that might distinguish them better (for instance, that of attention or visuospatial function). A useful model might be the High Sensitivity Cognitive Screen (Faust and Fogel 1989); although this was designed for use in non-delirious subjects with subtle impairments, it might well be useful in detecting states of mild or early delirium. Trzepacz *et al.* (1988) have reported encouraging preliminary results with a simple bedside test for delirium.

Secondly, the fluctuation of cognition that appears so important in distinguishing between delirium and dementia is not tested by these measures: none of them have, for instance, any items which are repeated. Test–retest reliability needs to be reconsidered as a useful criterion when the condition under study is defined as fluctuating and/or transient. The Saskatoon Delirium Checklist (Miller *et al.* 1988) is an attempt to embody the DSM-III criteria in a structured assessment, in which an observer (presumably medically qualified) rates the frequency of eight categories of impairment, including fluctuation. Also included is an assessment of the likelihood of a physical cause for the symptoms, but this item accounts for only four points out of a possible score of 40, so a high score on this instrument is possible without satisfying criterion E for DSM-III Delirium (see above). CAMDEX (Roth *et al.* 1988) is a standardized schedule designed primarily for the diagnosis and assessment of dementia, but it also contains items to enable differential diagnosis from other organic and functional disorders, including 'Clouded/delirious states'. As well as an examination of cognitive function, CAMDEX includes an informant interview in which any 'confusion' is rated in terms of its onset, duration and fluctuation. However, given the CAMDEX operationalized diagnostic criteria for 'Clouded/delirious states' (Table 1.5), it is difficult to see how this instrument can distinguish between dementia and delirium, since the criteria are met by any subject who is disoriented in time and place, and who shows impairment of recent memory.

Although there are considerable methodological difficulties associated

Table 1.5 CAMDEX diagnostic criteria for clouded/delirious state

Any two of the following:

A: Change in level of awareness manifest in any three of the following:

    (1) slowness and vagueness in thought;

    (2) markedly impaired ability to focus and sustain attention and concentration;

    (3) faulty comprehension with misinterpretation of surroundings;

    (4) periodic excitement or stupor;

    (5) incomplete arousal with periodic drowsiness.

B: Rapid onset of change in level of consciousness and cognitive function present for less than six months.

C: Disorientation in two of the following:

    (1) time;
    (2) place of abode;
    (3) place at present time.

D: Impairment of recent memory (defects in recording or recall of recent events).

E: Visual (more rarely auditory) perceptual distortion in the form of illusions or hallucinations of a fearful quality and often delusions of an anxiety laden or other persecutory nature.

F: Marked fluctuation in level of consciousness and cognitive performance.

with the use of the MMSE and other standardized assessments of cognitive dysfunction in the definition of delirium, they nevertheless represent a great advance on the unstandardized and idiosyncratic clinical diagnoses that have been employed in most investigations of this subject.

'The elderly'

So far, we have not considered the second part of the title of this book. The reasons for devoting this account of delirium to those of advanced years are as follows:

1. Delirium is more common in older patients than in younger ones. Studies of medical and surgical samples have consistently found higher rates in the elderly; Doty (1946) reported that the rate of delirium was increased by four times in patients aged over 40 years, and the highest rate found by Willi (1966) was in patients aged over 70 years. This association is probably due to age-related factors such as brain disease, altered drug metabolism, and sensory impairment; there is no evidence that 'old age' of itself predisposes to delirium, as is often implied by general reviews of the subject (Lipowski 1980b; Davison 1989).

Not all studies show an association between delirium and advanced age. In a recent investigation of 133 consecutive acute medical admissions, Cameron et al. (1987) found no significant age differences between delirious and non-delirious patients. Their list of the medical problems associated with delirium in their sample suggests why they found no difference: the two youngest delirious patients were in the terminal stage of AIDS. As the incidence of this particular cause of delirium increases in the young adult population, it will probably upset the traditional association between delirium and old age.

2. Those over retirement age in most developed societies are still subject to gross ageist prejudices and generalizations, and as we will argue for greater awareness and a more positive attitude to the remediable aspects of disorder in this age group, it is better to avoid confusing these arguments with evidence, say, from delirium in childhood.

For ease of understanding, 'the elderly' are defined here as all those over 65 years of age. Although this chronological definition is commonly used to separate the elderly from other adults, it should be noted that it is probably inadequate in several respects. First, the cut-off age of 65 years may no longer be a good one for discriminating between groups of the population with markedly different experiences of disorder (those aged 65 to 70 years may nowadays be very little different from those aged 55 to 60). Second, the concept of chronological, as opposed to physiological, emotional or cognitive age may be misleadingly simplistic. Third, there are also differences between the experience of old age (in nutritional, material, physical, psychological and social terms) that make comparisons across cultures difficult, even in the same time period. However, these objections have not yet been very much evident in the published literature on delirium, and will not be considered further in this context.

The importance of delirium in the elderly

Traditionally, some conditions demand more attention than others because of their frequency, distress and disturbance of functioning, and the associated mortality, treatability, preventability, or political impact (Macdonald *et al.* 1989). Disorders vie for attention with each other on different grounds – for instance the frequency and economic effects of lower back pain against the fatality and social threat of AIDS. Therefore, in order to assert the importance of delirium in the elderly, its epidemiology is reviewed here in terms of the frequency, the distress caused by it, and the outcome. The treatability is dealt with in Chapter 5.

The size of the problem

Medical samples

Naturally, determinations of frequency are heavily influenced by the population chosen for study, the definition of the condition, and the means chosen to assess the patient. Robinson (1956) found delirium in 40 per cent of patients over 60 years in a neurological ward, and in another early attempt to judge the size of the problem, Bedford (1959) estimated the prevalence of 'acute confusional states' in elderly patients admitted to the Oxford Geriatric Unit as high as 80 per cent, the highest prevalence rate ever reported. In contrast, Simon and Cahan (1963) estimated that only 13 per cent of hospital patients over 60 years with no previous history of dementia had delirium, and most subsequent studies of elderly medical patients have reported prevalence rates of this order.

In a study combining the mental test scores carried out by an unspecified number of interviewers with some details of cognitive decline in 588 patients aged 65 years and older admitted to 21 geriatric units throughout the United Kingdom, Hodkinson (1973) found that about 25 per cent of newly admitted in-patients satisfied his criteria for 'confusion'. These criteria were a Mental Test Score of 24 or less and a decline in less than two weeks, having been cognitively intact three months previously. 'Confusion' was distinguished from dementia by course alone, and occured in 44 per cent. The proportion remaining in hospital more than one month was the same (40 per cent) for patients defined as 'normal' as for those deemed 'confused', but was 60 per cent of those categorized as dementia. Some of the latter improved, and he suggested that these had both dementia and confusion. Various physical conditions were listed as causes of 'confusion', but he also included 'depression', which accounted for 12 per cent of cases. He regarded these patients as either performing badly on cognitive testing because of depression, or actually rendered delirious by tricyclic antidepressants.

In a series of 130 unselected patients aged over 65 years admitted to an acute medical ward, Bergmann and Eastham (1974) found 12.3 per cent suffered from acute delirious states (16 per cent of those not excluded for various reasons, including a past psychiatric history). They used an unstructured clinical assessment, based on a glossary with a mutually exclusive hierarchical structure as their diagnostic framework (organic states were at the top), and excluded 30 (23 per cent) of the sample for various reasons. No account was taken of subsequent course or outcome in the diagnosis. The prevalence of 'senile psychosis' was only 7 per cent – a far cry from the prevalence of dementia in new admissions to general medical wards today, and markedly different from Hodkinson's findings for geriatric wards at about the same time.

Seymour et al. (1980) studied all patients of over 70 years admitted as an emergency to three general medical wards. They found a 16 per cent prevalence of delirium in 71 patients using a definition based on either subsequent changes in a simple orientation and memory score, or on the history of cognitive decline.

The contribution of Anthony et al. (1982) to the methodology of the assessment of delirium has already been discussed (page 11). The result of the prevalence study was that nine (9.3 per cent) were judged to have a 'pure' delirium, and one (1 per cent) was delirious and demented. However, the ages of these patients ranged from 20 to over 80 years.

In a study of 2000 consecutive medical admissions aged 55 years and older, Erkinjuntti et al. (1986) found that 15.1 per cent were delirious on admission. In this study delirium was defined by DSM-III criteria. There was a significant association with increased age, and 24.9 per cent of the delirious patients were also demented. Another recent study by Rockwood (1989) examined 80 patients, ranging in age from 65 to 91 years, admitted to general medical wards in Canada, and found a prevalence of 25 per cent of 'acute confusion' using DSM-III criteria for delirium. He inter- viewed each patient daily (except at weekends) using both his own rating scale and that of Reisburg et al. (1982) for dementia. There was a non- significant trend for the delirious patients to have a longer stay in hospital, and three of the four patients who died had delirium. Delirium was a transient condition in his sample, with 75 per cent of cases lasting less than 24 hours and none lasting for more than a week.

Surgical samples

Elderly surgical patients seem particularly prone to delirium (Whitaker 1989). Cardiothoracic surgery has traditionally been associated with a particularly high rate of post-operative delirium, and this phenomenon has been the subject of many studies. In one series of older patients undergoing open heart surgery (mean age 53 years), the prevalence of postoperative delirium was 79 per cent (Shaw *et al.* 1986). In a recent meta-analysis of 44 such studies of post-cardiotomy delirium conducted between 1963 and 1987, Smith and Dimsdale (1989) have examined the trend in prevalence rate over this period, and have sought to identify associations with other factors that are consistent across the studies. They point out that it is difficult to make detailed comparisons between studies, since there is no consensus as to the definition of delirium or onset time, and the factors studied vary. On the basis of their meta-analysis, they found a relatively constant prevalence over time of 32 per cent; there was no evidence of a significant change between the periods 1963–1974 and 1975–1987. Among the factors that showed no correlation with post-cardiotomy delirium were sex, past psychiatric history, intelligence, and time on by-pass support systems or in intensive care units. There was a slight correlation with age, but the highest correlation (-0.60) was with pre-operative psychiatric intervention, suggesting that this may have a preventive effect.

Hip replacement following fractured neck of femur is another procedure with a high rate of post-operative delirium in the elderly. The prevalence of delirium following hip replacement was 32 per cent in the series of patients studied by Furstenberg and Mezey (1987), and 35 per cent in a similar series reported by Berggren *et al.* (1987); in a more recent series reported by these authors the prevalence was 61 per cent (Gustafson *et al.* 1988). There is a correlation between falls, functional psychiatric disorder, and psychotropic medication such as tricyclic antidepressants and hypnotics (Berggren *et al.* 1987; Vetter and Ford 1989); it may be that this medication contributes both to the fall and to the subsequent post-operative delirium (Lipowski 1989).

General and elective surgery in the elderly appears to carry a lower risk of post-operative delirium. Millar (1981) examined 101 patients aged 65 years and older both before and after elective surgery, using the Clinical Interview Schedule, an instrument designed for the assessment of minor psychiatric morbidity in community settings (Goldberg *et al.* 1970), supplemented by a memory and orientation scale. The mean age of the sample was 73 years. Incorporating nurse interviews and nursing records in an unclear way, he estimated that 9 per cent had new evidence of cognitive decline post-operatively, and a further 5 per cent had evidence of previous impairment as well. He regarded the majority (65 per cent) as 'quietly' impaired, whilst the remainder presented behavioural problems such as hitting out, verbal abuse, and pulling out intravenous lines. There was no attempt to distinguish delirium from dementia. Of those cognitively impaired, 70 per cent improved over four days, 15 per cent died and 15 per cent remained impaired over a week. He found little correlation between post-operative cognitive impairment and sensory impairment, but felt there was a definite correlation with severity of physical illness.

Psychiatric samples

Golinger (1986) reviewed the case-notes (charts) of all patients referred to a psychiatric consultation-liaison service from a surgical department over two years. Thirty-one of the 150 patients (20.7 per cent) satisfied DSM-III criteria for delirium. He suggested that the prevalence was related to age, since 9.8 per cent of the patients under 60 years and 43.8 per cent of the patients over 60 years were delirious. The mean age of the sample was 47 years, and the reasons for referral were not discussed. Two per cent of his sample suffered from dementia.

Sirois (1988) examined the case records (charts) of 793 referrals to a psychiatric liaison-consultation service in a general hospital, and estimated that 12.6 per cent satisfied DSM-III criteria for delirium, a disproportionate number of whom were men. The cases were referred because of 'confusion with incoherent speech, agitation and disorientation', so it is likely that quietly delirious patients were not referred, and that this figure is an under-estimate.

In a study of new admissions to a psychogeriatric unit, Koponen et al. (1989b) identified 70/523 (13.4 per cent) who met DSM-III criteria for delirium (mean age 75 years). An unknown number of cases with delirium caused by alcohol withdrawal were excluded. Quiet delirium was probably under-represented in this sample, since most cases were admitted from other medical units where their treatment had been complicated by disturbed behaviour associated with the delirium. Evidence of possibly predisposing structural brain disease was found in 81 per cent of cases.

There have been few reports of the prevalence of delirium in non-hospitalized elderly people. In community surveys, delirious subjects are likely to be removed from the sample frame by death or institutionalization (Jacoby and Bergmann 1986). One particular group that is probably at risk of increased rates of delirium is the chronically ill and functionally dependent population resident in geriatric nursing homes. In one study of psychiatric disorder in a sample of 50 residents of a nursing home in the United States, three (6 per cent) were found to be delirious according to DSM-III criteria as a result of drug intoxication (Rovner et al. 1986). In a consecutive series of subjects referred to an on-site psychiatric liaison-consultation service to a nursing home in the United States, 6 per cent were diagnosed as delirious (criteria not stated) (Bienenfeld and Wheeler 1989).

Delirium and distress

It is extremely unlikely that delirium is a pleasant experience. Cutting (1980, 1987) has recounted the experiences of 74 consecutive patients referred to a psychiatric liaison service (mean age 57.4 years) from medical wards with what he termed 'psychosis' secondary to some clearly identifiable organic cause. Concentrating on delusions, he summarized their experience as that of being helpless onlookers while some atrocity or misdeed was enacted, with a nightmare-like quality. For instance a patient of 69 years with 'electrolyte imbalance' believed that her husband had been eaten by a cannibal, and that drunken nurses on the ward were doing unspeakable things for money, and that she had venereal disease. The perceptual abnormalities included things crawling over the patient, things becoming enmeshed in the patient's body, the bed moving, babies crying, and relatives calling for help. Although this series of patients may have been referred precisely because of their distress (or that of the staff looking after them), it is likely that the unreferred 'quiet' delirious patients may have had similar experiences. How much of this distress is remembered by the patient is unclear; while amnesia for the period of the delirium is commonly reported, the report of post-traumatic stress disorders following delirium (Mackenzie and Popkin 1980) suggests that the disturbing experiences are sometimes remembered. In some cases, delirious experiences may not be reported because patients are afraid that they will be thought mad (Heller et al. 1970). We know of one patient who, though unable to recall anything of a post-cardiac arrest delirium (or, indeed, her stay in hospital at all) refused to visit the ward on which she had been delirious, for reasons she could not express.

The outcome of delirium

Mortality

The outcome for delirious states has always been regarded as that of a 'crisis' – kill or cure. Traditionally, delirium was regarded as the harbinger of death, as Shakespeare describes in the case of Falstaff: 'For after I saw him fumble with the sheets, and play with flowers, and smile upon his finger ends, I knew there was but one way; for his nose was sharp as a pen, and 'a babbled of green fields . . .' (Henry V, II. iii). Excess mortality can be seen in two ways: either the tendency to delirium is proportional to the severity of the physical illness, the latter being the cause of the mortality, or the presence of delirium actually worsens the outlook independently of any physical cause of mortality – for instance by making management of the physical illness more difficult in a number of ways.

This issue is discussed in Chapter 5. Whatever the mechanism, the excess mortality in delirious states is now well established. The different rates reported reflect, as ever, the differing ages and other selection artefacts in each study.

Bedford (1959) reported that one third of patients exhibiting delirium died within a month. Similarly, the mortality in the patients studied by Bergmann and Eastham (1974) was 37.5 per cent. Rabins and Folstein (1982) found that medically ill patients referred for psychiatric assessment who were found to be delirious (using the MMSE) had mortality rates similar to those found by Roth 27 years before (1955) – 40 per cent were dead by 6 months and 50 per cent by two years. Hodkinson (1973) found that one quarter of delirious elderly patients died within a month; a mortality rate that was about twice as high as that for a comparable non-delirious group. Patients with the most severe cognitive impairment had the poorest outcome.

The outcome of delirium among those who survive appears to be good. Of the survivors at one month in Bedford's sample (1959), 80 per cent recovered in less than a month, and only 5 per cent remained confused for more than six months. Thirty-five per cent of the delirious group in Hodkinson's 1973 survey were discharged within one month, compared with 21 per cent of the demented patients. Similarly, the discharge rate in Bergmann and Eastham's study (1974) was 44 per cent. In the series reported by Rockwood (1989), over 80 per cent recovered. Since much of the delirium experienced by elderly patients is mild, transient, and undetected, these reported recovery and discharge rates are probably underestimates.

Length of hospital stay

In these cost-conscious times, the financial burden imposed by physical and psychiatric disorders upon the health services is an important consideration when making judgements about their importance. Delirium is associated with increased length of hospital stay. In their sample of elderly people with hip fractures, Berggren *et al.* (1987) found that those with post-operative delirium had a mean hospital stay of 77 days compared with a mean of 22 days in the non-delirious group. Similarly, in a prospective study of 133 medical patients admitted to hospital in the United States, the mean length of stay of those who developed delirium was 21.6 days compared with 10.6 days for the non-delirious patients (Thomas *et al.* 1988). The delirium itself appears to be relatively transient in many cases (Rockwood 1989), so it may be that patients with delirium are more severely physically ill and require longer periods in hospital for that reason; further studies are needed to determine the extent to which prolonged hospital stay is specifically due to delirium and its consequences.

Chronic psychiatric disorder

There is a persistent suggestion in the literature that there is a transition from delirium to dementia in a proportion of cases (Lipowski 1983). Evidence on this point is sparse, but it appears that any enduring cognitive impairment or cognitive decline following an episode of delirium in the elderly is most probably caused by a pre-existing dementia, or an acute cerebral insult such as a cerebrovascular accident or peri-operative hypoxia that precipitated both the delirium and the persistent defect. In a follow-up assessment of their sample of delirious psychogeriatric patients, Koponen *et al.* (1989*b*) found that significant decline in cognitive function as measured by the MMSE at one year follow-up occurred in one third of cases. This decline was associated specifically with the addititional diagnosis of probable Alzheimer's disease or multi-infarct dementia on admission; subjects without evidence of central nervous system disease on the index admission showed no significant decline in their MMSE scores at follow-up.

Functional psychosis has also been reported as a rare outcome of delirium (Bleuler *et al.* 1966; Lipowski 1985). The study of Koponen *et al.* (1989*b*) suggests that in the elderly this transition may be due to a common underlying disorder, since the only patient who progressed from delirium to a functional psychosis was also severely hyperthyroid. However, elderly people with severe depression and mania are at increased risk of developing delirium (see page 39), and the transition to functional psychoses may simply represent the emergence of these disorders following treatment of a supervening delirium due to associated self-neglect or exhaustion. In cases where the delirious delusions persist and become systematized following recovery, it is possible that the delirium has precipitated a paranoid disorder.

2 Clinical assessment and diagnosis

Those that be frentick have a continuall fever, & be madde, for the most part they cannot sleep. Sometimes they have troublesome sleepes, so that they ryse up, & leap, & crie out furiously, they babble words without order or sense, being asked a question, they aunswere not directly, or at the least rashly, and that with a loud voice, especially if you speake gently to them.

Barrough (1583)

At all ages, the diagnostic process in relation to delirium has two stages: first, it is necessary to establish that the patient is delirious, and second, the underlying cause of the delirium must be identified. This chapter focuses on the first of these stages, the preliminary assessment, investigation and differential diagnosis of delirium in the elderly.

Assessment

Abnormalities in every aspect of the mental state have been described in association with delirium. Elderly people with delirium will usually show some of the central features of the syndrome, but it is important to bear in mind that the clinical picture may be less complete and characteristic than it is in younger people. For example, unlike the florid disturbances that often accompany delirium in younger adults, delirium in the elderly may be quiet and easily overlooked by those who do not know the person well (Millar 1981). For this reason, the clinical assessment of delirium in the elderly requires both a detailed history and careful examination and review of the mental state.

History

The patient's own account of events will be nearly always be unreliable or non-existent, and a collateral history from other sources is mandatory in all cases. In a community setting, such as the patient's own home, it is usually possible to obtain an account of the onset and progress of the disorder from relatives, neighbours, home helps and other regular contacts. In hospital, visiting relatives should be interviewed, and other key informants such as the GP should also be contacted. The nursing staff (particularly those on night duty) will often provide very useful information about a patient's current mental state. A reliable account of the mode of onset is of particular importance. Typically, delirium has an abrupt onset, over hours or days. In the elderly it may develop more insidiously in association with certain causes, such as chronic subdural haematoma or a slowly developing drug toxicity, but the history will rarely be more than a few months. Information about the course of the episode should always be sought. Diurnal fluctuations in severity with lucid intervals and nocturnal deterioration are highly indicative of delirium. If a history of any past or current physical or psychiatric illness can be obtained from relatives, or from the family doctor, this will also be of help in establishing the diagnosis, as will information about current medication and recent changes to this.

Case 1

An 85-year old man was referred by his general practitioner to the psychogeriatric service with a five-day history of complaining that men had pulled him from his home at night, had punched him, and were waiting for him outside. He had become restless and fearful, quite different from his normal, placid self. There was, however, no history of sleep disturbance. The general practitioner had examined him and, apart from mild hypertension which had been well controlled with a beta-blocker and diuretic for some time, no underlying acute physical cause had been discovered, and a psychiatric condition was therefore suspected.

The psychogeriatric team obtained from his wife the history of an inexorable decline in his memory and concentration over the previous seven years, but little change in daily functioning, since she had always done everything for him. He had not been incontinent before, and could previously dress himself. A few weeks previously, he had been to visit relatives in the United States, and while there had developed a chest infection and been admitted to hospital. Whilst in hospital his regular antihypertensive medication had apparently been regarded as hopelessly obsolete, and had been changed to an angiotension converting enzyme (ACE) inhibitor only available in the United States. He had been discharged improved, had returned to Britain and was well until the supply of this medication was exhausted. He was then prescribed a different ACE inhibitor available in Britain, and two days later became delirious as described. Three days after reverting to his original antihypertensive medication he returned to his normal self, with continuing evidence of moderate cognitive impairment in clear consciousness.

Informants should be asked about any similar episodes in the past, their cause, treatment and outcome. In elderly delirious patients it is important to determine the extent of any pre-existing cognitive deficit, since it is from this prior level of cognitive function that the recent decrement must be assessed. Table 2.1 is a checklist of the key questions that should be asked of an informant.

If the patient lives alone and there are no helpful informants, a brief

Table 2.1   The informant history

How long have you known [the patient]?

When did you last see [the patient] before admission?

How was he/she then?

Do you find [the patient] now confused, or muddled, or forgetful?

IF YES: Can you say when this first became apparent to you? When was the first time you noticed [the patient] was confused / muddled / forgetful?

When, to the best of your knowledge, was [the patient] not confused or muddled at all?

Did it happen suddenly? Over what period (hours, days, years)?

In your opinion, did anything happen to bring about this change? (a fall / an illness / any stress?)

How has [the patient] slept over the past 48 hours? Has [the patient] been drowsy during the day over this period? All of the time? Some of the time? Can [the patient] be woken / roused?

Have you noticed any of the following:

    plucking at the bedclothes
    misidentifying strangers as familiar
    failing to recognise familiar relatives and friends
    inexplicable calling-out
    seeing things that aren't there
    unfounded accusations
    sudden mood changes or crying
    unusual physical aggression

---

examination of their domestic environment often provides a wealth of useful background information. Severe cognitive impairment in the context of a reasonably clean and well-ordered household is suggestive of uncomplicated delirium. By contrast, sticky carpets, mouldy food in the refrigerator, burnt kettles and an accumulation of rubbish are indicative of long-standing difficulties due to dementia or, rarely, severe depression. Assessment at home also provides an opportunity for accurate estimates of the patient's diet, medication history (prescribed and non-prescribed, hoarded and new), and alcohol intake.

Examination of the mental state

When taking a history from the patients themselves, attention should be paid to how recent events are recalled, sequenced, and made sense of, since this provides a good opportunity to make an unobtrusive assessment of alertness, concentration, memory, and thinking. Where the delirium is severe, formal assessment of the mental state may be impossible, but in moderate and doubtful cases brief cognitive function tests are helpful, particularly those that are sensitive to attention deficits, such as tests of temporo-spatial orientation and arithmetical function (serial sevens, digit span). When testing memory function in delirium, it is new learning ability that is most important; the patient should be asked to repeat the test material (a sentence, a name and address) both immediately after a single hearing and 3–5 minutes later. Language function may be briefly tested by asking the subject to name objects (production) and obey commands (comprehension). Visuo-spatial function is commonly disturbed in delirium, and may be tested by asking the subject to draw a clock and copy a diagram. Other brief bedside tests of cortical function may be diagnostically valuable; for example, persistent errors in the face–hand test in which the subject is asked to report the locations of simultaneous light touches to the cheek and hand strongly suggest organic rather than functional impairment. Fink *et al.* (1952) found that errors on the face–hand test were greatest in those whose organic disorder was of rapid onset and short duration. The validity of this test in the elderly has been demonstrated by Kahn *et al.* (1960).

The role of formal psychometric testing in the diagnosis and assessment of delirium is limited, since all such tests require a good degree of co-operation and attention on the part of the patient, and this is usually lacking. Vigilance tasks such as simple reaction time to randomly presented signals measure the patient's ability to monitor stimuli over time, and may be useful in demonstrating the presence of mild and fluctuating disturbances of alertness and attention in co-operative subjects. Brief cognitive function tests such as the Mini-Mental State Examination (Folstein *et al.* 1975) typically test orientation in time and place, verifiable personal information, current general information, personal and non-personal remote memory, new learning ability and psychomotor skills (Table 2.2). Delirious patients score positively, but it is not possible to distinguish between delirium and other causes of cognitive impairment, such as dementia, on the basis of these tests, so their diagnostic value is limited (see page 11).

Table 2.2   The mini-mental state examination
(adapted from Folstein *et al.* 1975)

ORIENTATION	POINTS
1) What is the: year?	1
season?	1
date?	1
day?	1
month?	1
2) Where are we? country / state	1
county	1
town / city	1
hospital (street)	1
floor (house number)	1

REGISTRATION

3) Name three objects, taking one second to say each. Then ask the subject all three after you have said them. Give one point for each correct answer.   3

Repeat the answers until the subject learns all three.

ATTENTION AND CALCULATION

4) Serial sevens. Stop after five answers. Give one point for each correct answer.   5
(ALTERNATIVE: Spell WORLD backwards.)

RECALL

5) Ask for names of three objects learned in question 3. Give one point for each correct answer   3

LANGUAGE   2

6) Point to a pencil and watch. Have the subject name them as you point.

7) Have the subject repeat 'No ifs ands or buts'.   1

8) Have the subject follow a three-stage command: 'Take the paper in your right hand. Fold the paper in half. Put the paper on the floor.'   3

9) Have the subject read and obey the following: 'CLOSE YOUR EYES'   1

LANGUAGE *(cont.)*	POINTS
10) Have the subject write a sentence of his or her choice. The sentence should contain a subject and an object, and should make sense. Spelling errors are ignored when scoring.	1
11) Have the subject copy the following design:	

All sides and angles must be preserved, and the intersecting sides form a quadrilateral.   1

MAXIMUM SCORE   30

Clinical investigations

Electroencephalography (EEG) is of diagnostic use in mild and doubtful cases of delirium (Engel and Romano 1959). In delirium secondary to metabolic disorders and brain lesions there is progressive slowing of the alpha rhythm, together with a reduction of its voltage. When this slowing reaches 5 cycles per second the alpha rhythm is no longer disrupted by eye opening. With increasing severity the EEG becomes dominated by asynchronous runs of 5–7 cycles per second theta activity, and these slow waves become generalized as the delirium becomes more severe. High voltage delta waves appear (Bickford and Butt 1955), which tend to be frontal and bilaterally synchronous (Fig. 2.1). These may not occur spontaneously in mild delirium, but can be induced by hyperventilation. There is a close correlation between the degree of slowing and the level of functional disturbance of attention, alertness, comprehension and memory in delirium (Romano and Engel 1944; Koponen *et al.* 1989*e*), so serial EEGs are useful for monitoring progress (Pro and Wells 1977). However, EEG slowing does not change in association with the nocturnal worsening of delirium unless there is an associated worsening of the underlying disorder. Generalized slowing of the alpha rhythm also occurs with advancing age in normal people (Obrist *et al.* 1963; Marsh and Thompson 1977), and in

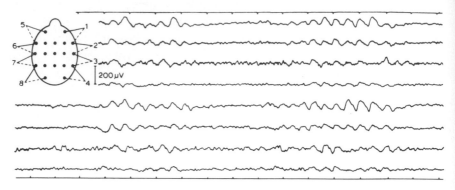

Fig. 2.1 Delirium associated with bronchopneumonia in a 60-year old woman. EEG shows bilaterally synchronous frontally predominant 2 Hz waves (Kiloh *et al.* 1981).

association with dementia (Obrist 1979), so the diagnostic value of the EEG is more limited in the elderly than in younger age groups (Lipowski 1983). As a rule, evidence of slowing relative to previously observed rates is a more significant indicator of delirium than the absolute rate.

Although studies indicate that these EEG changes are not specific for particular systemic metabolic disturbances (Dongier 1974), certain features are more pronounced in some conditions; for example the triphasic waves associated with hepatic encephalopathy and the high-amplitude sharp waves seen in herpes simplex encephalitis. Abnormalities such as spikes, sharp waves and wave-and-spike combinations usually suggest an intra-cranial cause for the delirium, particularly if these and other EEG abnormalities are localized (Obrecht *et al.* 1979).

Not all patients with delirium show the EEG changes associated with metabolic disturbance; in particular, during withdrawal states such as delirium tremens the EEG shows paroxysmal bursts of low voltage fast activity rather than slowing. In toxic delirium associated with drugs that reduce the convulsive threshold, such as chlorpromazine, the EEG may show drug-specific patterns of fast-wave activity (von Sweden and Mellerio 1988).

Another method of electrophysiological assessment, the visual evoked potential (VEP), may offer a simpler and quicker means of identifying and monitoring delirium. Zeneroli *et al.* (1984) have demonstrated that the VEP is altered in hepatic encephalopathy, with waveform changes and increased N3 latency. In a group of non-elderly adult patients, the degree of abnormality correlated with the severity of the encephalopathy, and abnormalities were apparent in pre-clinical cases. It has yet to be shown that VEP is a useful tool in the assessment of delirium in old age, where VEP abnormalities associated with dementia may complicate the picture (Coben *et al.* 1983).

Clinical features

In the prodromal phase of delirium the patient may complain of muddled thinking, and there is increasing difficulty with concentrating and judging the passage of time. The patient becomes irritable and restless, or lethargic and withdrawn. Mood disturbance is common, particularly if the patient is aware that something is the matter. Hypersensitivity to sensory stimuli may be reported, as may perceptual distortions, or effort in maintaining normal perceptions. The sleep–wake cycle often shows early signs of disturbance, with insomnia by night and drowsiness by day, vivid dreams and hypnogogic and hypnopompic hallucinations. In an attempt to maintain a façade of normality in the face of cognitive testing, the subject may be evasive, or refuse to co-operate, or respond with a catastrophic reaction when pressed (Beresin 1988). In demented elderly people, the onset of a delirium may be heralded by sudden changes in behaviour, or sudden deterioration in the levels of self-care capacity, mobility, or urinary continence. Sometimes delirium represents an emergence from unconsciousness rather than a falling away from full consciousness, and in these cases it will be preceded by episodes of coma or stupor (Lipowski 1980*a*).

Cognitive function

Central features of delirium are a reduced ability to focus, sustain and shift attention to external stimuli, and disorganized thinking, and these manifest themselves clinically as global cognitive impairment. There is a rapidly developing disorientation in time and place, together with disorientation in person if the delirium is severe. The attentional deficit results in impaired concentration and distractibility. Typically, this cognitive impairment fluctuates over time, with lucid intervals in the morning and maximum disturbance at night when the patient is fatigued and sensory input is reduced.

Delirious patients often show disturbances or reversals of their sleep–wake cycle, with drowsiness by day and insomnia by night. Occasionally, patients are persistently hyperalert. Delirium often makes its first appearance as an episode of nocturnal disturbance, when the patient is awake, disoriented, hallucinating and agitated (Lipowski 1983). Conversely, the restoration of quiet sleep is often an early sign of improvement, and in some cases the delirium may resolve with a prolonged 'terminal sleep'.

Thinking

In delirium, thinking is progessively disturbed. If the delirium is mild, the only abnormality may be a slowing down or speeding up of the stream of thought, but as it becomes more severe the capacity to make judgements, to grasp abstract concepts and to reason logically are all impaired, and thinking eventually becomes undirected, disorganized and incoherent. The content of thought is also disturbed in delirium. As the patient's awareness of the external environment diminishes, so the significance of the internal world increases, and the patient loses the ability to distinguish between them (Lishman 1987). False perceptions and misinterpretations of the outside world combine with intrusive private fantasies and dreams, and this can result in the sudden and powerful formation of delusions. These delusions are usually described as persecutory, but Cutting (1987) found that organically induced delusions were more accurately characterized as belief in 'imminent misadventure to others' and in 'bizarre happenings in the immediate vicinity'. Compared with schizophrenic delusions, organic delusions in his study were more likely to involve others as the victims of the imagined drama. Another delusional belief sometimes reported by delirious subjects is that friends and/or relations have been replaced by identical imposters – the so-called Capgras syndrome. Various authors have commented on the possible organic basis for this syndrome, and Christolodou (1977) has suggested that it is due to pathological changes in the non-dominant occipital lobe.

Delirious delusions are for the most part transient, poorly systematized, inconsistent, and stimulus-bound; very occasionally, they may persist and become more systematized following recovery from the delirium. The following case describes an individual whose delusions about the nursing staff persisted for at least a year after complete recovery from the delirium.

Case 2

An 80-year old widow with a long history of chronic obstructive airways disease was admitted in respiratory distress to a medical ward. She complained that, one night, the nursing staff substituted her notes (chart) for those of another person, who had been or was a prostitute. These notes were discussed loudly by the nurses on night duty, and shown to portering staff who came on to the ward, and then to other patients. She knew that they were not her real notes, as the nurses referred to 'blue eyes' (hers were brown), and the allegation of being a prostitute was unfounded. She tried to stop them reading out passages from the notes, and became distressed. She remained convinced that these events continued each night until her discharge. After discharge, she continued to allege that the notes were not her own, and refused to return to that hospital when her chest condition deteriorated. She was referred to the local psychogeriatric service one year after the onset of the delusion, and no significant cognitive impairment or other psychiatric disorder was discovered. The delusion was held in clear consciousness. Because of her physical state and the encapsulated and specific nature of the delusion, no psychotropic medication was prescribed, and the patient's delusion was abolished by being allowed to peruse her notes with the psychiatrist. This reassurance was only temporary, however; a subsequent admission and further episode of delirium restored the delusion to its former fixed state.

According to Lipowski (1983), the thought content of elderly delirious patients is likely to be more impoverished and stereotyped than that of younger adults. This may be so for patients with pre-existing dementia (Beresin 1988), but there is no evidence to suggest that non-demented elderly patients are less vulnerable to the formation and expression of delirious delusions; in one study of geriatric in-patients with delirium, approximately 50 per cent expressed either 'persecutory delusions' or 'unrealistic suspiciousness' (Simon and Cahan 1963).

The speech and writing of delirious patients reflect the form and content of their thought, and provide valuable evidence of disturbance. Speech may be slow or pressured, there may be difficulty in finding words and naming objects, with paraphasias and circumlocution. In a study of patients undergoing barbituate anaesthesia, Chedru and Geschwind (1972a) described sensory dysphasia in the period of delirium preceding unconsciousness. Spontaneous utterances tend to be stereotyped, perseverated and eventually incoherent. Sometimes subjects may become mute. The speech volume ranges from whispered muttering to shouting and screaming. Where English is the second language, there may be a return to the native tongue. Decrements in writing ability also occur in delirium; Chedru and Geschwind (1972b) found that dysgraphia and errors in grammar and spelling were all sensitive indicators of delirium in their subjects.

Memory

In delirium the registration, retention and recall of experience is disturbed. In particular, the impaired registration results in new learning deficits and anterograde amnesic gaps, and these provide sensitive clinical indicators of mild and early delirium (Lishman 1987). There is usually a partial or complete amnesia for the period of the delirium once it has remitted, but certain psychotic experiences may be recalled. Impairment of remote memory in delirium is less common, but recall is inevitably disturbed in more severe cases. Sometimes the delirious patient's responses to questioning, while incorrect, nevertheless indicate that some new learning has taken place since the onset of the disorder (Geschwind 1982). The phenomenon of reduplicative paramnesia, in which the patient holds incompatible beliefs, often about spatial orientation, without being aware of the inconsistency, has been described in delirium (Patterson and Zangwill 1944; Geschwind 1982), and false memories and confabulation may also be prominent in some cases (Wolff and Curran 1935). There is a tendency to mistake the unfamiliar for the familiar, rather than the other way round (Levin 1956). This process has been explained in psychodynamic terms as the subject's defence against the experiential chaos of delirium (Beresin 1988), but inappropriate experiences of familiarity such as déjà vu are known to occur in temporal lobe epilepsy, and their occurrence in delirium may also be the result of physiological dysfunction of the temporal lobes.

Perception

A wide range of perceptual distortions, illusions and hallucinations have been described in association with delirium. They occur in all sensory modalities, but visual, auditory and visual-plus-auditory disturbances are the most common. They are a striking aspect of the mental state in delirium, and their diagnostic importance was emphasized by early authors such as Kraeplin (1889) and Bonhoeffer (1910). As with most delirious phenomena, the reported prevalence of hallucinations in elderly patients with delirium varies considerably across study samples, probably as a function of the methods used to collect the data; using a systematic interview, Aggernaes and Myschetzky (1976) found that most of their patients had experienced hallucinations. Although hallucinations are common in delirium, they are not a core feature of the syndrome, and may occur in the absence of delirium or any other psychiatric disorder. In the elderly, vivid, elaborate visual hallucinations, often of children or miniature adults, occasionally occur as an isolated psychopathological phenomenon associated frequently with pathology of the visual system – the so-called Charles Bonnet syndrome (Damas-Mora et al. 1982; Gold and Rabins 1989). In some of these cases the phenomena may be facilitated by episodes of transient, sub-clinical delirium (Berrios and Brook 1982). Insight into the hallucinatory nature of the experience is retained, and the subject typically does not complain for fear of being thought mad. Visual hallucinations and illusions are also common following bereavement (Rees 1971; Olson et al. 1985), and tend not to be reported for the same reason.

In general, perceptions in delirium are either unusually dull or unusually acute. Other reported distortions include metamorphopsias such as seeing objects as smaller than they are (micropsia) or larger than they are (macropsia), or otherwise distorted in shape or number (Willanger and

Klee 1966). Distortions of body image, weight and size have also been described (Lishman 1987). Illusions are the result of misinterpreting external stimuli, for example, seeing faces in the pattern of the wallpaper, or misrecognizing strangers as relatives. Hallucinations of all degrees of complexity occur in delirium, from simple flashing lights and noises to animals, people and voices. Patients may be observed in conversation with, or shouting at hallucinatory figures. Sometimes visual hallucinations are themselves distorted, as in Lilliputian hallucinations where people and animals are seen in miniature (Fish 1974). Autoscopy is a rare 'hallucinatory perception of one's own body image projected into external visual space' (Lukianowicz 1958) that may occur in delirium; according to Fish (1974), the association between autoscopy and physical disorders such as stroke or severe infectious disease accounts for the German folk-belief in the Doppelgänger as the harbinger of death. Tactile hallucinations are less common than visual and auditory hallucinations in delirious patients, and usually take the form of paraesthesiae and crawling sensations (Wolff and Curran 1935). Illusions and hallucinations occur more commonly when the sensory input is ambiguous or partial, for example at night, and in delirious elderly patients with pre-existing visual and auditory impairment.

Psychomotor behaviour

The delirious patient may be hyperactive, or hypoactive, or both, with behaviour oscillating rapidly and unpredictably between the two (Steinhart 1979; Lipowski 1983). Hypoactivity with reduced spontaneous and purposeful motor activity, and slowing, hesitation, and perseveration of movements and speech, is the form most commonly encountered in the elderly. It is not dramatic, and can easily be missed on a busy medical or surgical ward, particularly if the patient already has a degree of cognitive impairment due to dementia. Much more dramatic, if less common in this age group, is the hyperactive delirium associated with drug withdrawal states and certain systemic infections. In hyperactive delirium the disturbed behaviour usually reflects the patient's disturbed inner world of hallucinations and delusions, and is often accompanied by hyperalertness, mood disturbance, autonomic arousal and an exaggerated startle response. For the most part, overactivity in delirium is purposeless and repetitive; a good example is the persistent plucking and picking at bedclothes, which has been observed since ancient times and described as carphology (Hippocrates), crocydismos (Arateus), or more prosaically as 'wool-gathering'. Sometimes, delirious patients show more complex stereotyped movements, the so-called 'occupational delirium', since these stereotypies often represent habitual, work-related activities. Semi-purposeful overactivity in delirium is less common, and is usually inappropriate and clumsily performed; however, it is the resulting restless wandering, searching and above all the verbal and physical aggressive outbursts that cause the greatest management problems, particularly in hospital and residential environments. Violent behaviour is not common in elderly delirious patients; in the series reported by Simon and Cahan (1963) it occurred in only 10 per cent of cases. As a rule, concurrent physical debility and the poor co-ordination of aggression ensure that little harm comes to physically fit carers such as nursing staff and relatives, but other patients may be at risk from direct or inadvertent assault. The greatest risk is to the elderly delirious patients themselves, who may become exhausted, or fall, risking hip fractures and head injury (Lipowski 1980a).

Emotion

Delirium is usually accompanied by profound affective changes. When hypoactive, patients appear apathetic and show little affective responsiveness, but assessment of their subjective state is rarely possible and it may be that they are in fact significantly distressed. The distress associated with delirium is more apparent in hyperactive patients, who typically display fear, depression and rage in response to their hallucinations and imagined persecutions. Vocalizations such as moaning and screaming give a good indication of the emotional state. Prominent crying may also be a sign of delirium. In one prospective study of hospital patients referred to a psychiatric consultation-liaison service because of crying (Green et al. 1987), only 20 per cent had an uncomplicated psychiatric disorder; the remainder had either an organic disorder, a combination of psychiatric and organic disorders, or a previously undescribed entity of 'essential crying'. The most common organic disorder reported in these patients was 'bilateral hemispheric dysfunction' associated with dementia or delirium; the most common psychiatric disorder was major depression.

Profoundly distressed subjects may attempt to kill themselves, but fortunately delirium seems to interfere with the successful completion of suicide (Senbuehler and Goldstein 1977). Very occasionally certain psychotic experiences, such as Lilliputian hallucinations, elicit curiosity and pleasure rather than fear or anger in the delirious subject (Fish 1974). Fear is regarded as the predominant affect in delirium, but it has been reported that depression also occurs in up to 60 per cent of delirious geriatric patients (Beresin 1988). Not surprisingly, the patient's pre-morbid personality, mood state, life-events and relationships all have an impact on the nature and extent of affective responses in delirium (Wolff and Curran 1935; Lishman 1987).

Physical signs

Hyperactive and fearful patients are often autonomically aroused, with dilated pupils, tachycardia, dry mouth, diaphoresis, and elevated blood pressure. Neurological symptoms occuring in delirium include motor abnormalities such as tremor, myoclonus, ataxia, choreiform movements, asterixis (the 'flapping tremor' of the hyperextended hands first described in hepatic encephalopathy (Sherlock 1985)) and dysarthria, and cortical signs such as dysphasia and dyspraxia. Elderly patients may become incontinent of both urine and faeces when delirious. In all cases, physical symptoms due to the delirium must be distinguished from those caused by any underlying physical illness or intoxication.

Screening for delirium

Despite the prevalence of delirium in hospital settings such as casualty departments (emergency rooms) and in-patient units, information about its presence is hard to come by and doctors are often unaware of its existence, particularly in their elderly patients (Knights and Folstein 1977; McCartney and Palmateer 1985; Cameron *et al.* 1987). This is only partially due to the relatively quiet and atypical clinical picture of delirium in this age group. Modern hospital staff tend to be preoccupied with the techno-logical aspects of their practice, and overlook the cognitive and emotional problems in their patients.

Case 3

A 78-year old widow, living alone, had a three year history of sudden decrements in her mental state, with dysphasia, dyspraxia, and memory difficulties. She had slow atrial fibrillation and chronic leg oedema complicated by significant varicose veins, and she was taking digoxin, a 'potassium-sparing' diuretic and aspirin. She was functioning well at home – cooking (after a fashion), travelling by bus (becoming lost only occasionally), shopping, and attending church functions. She was admitted for varicose vein stripping and became more muddled pre-operatively. Pre-operative biochemistry was not checked. She suffered a profound bradycardia under anaesthetic, and was found to have toxic levels of digoxin and hypokalaemia. She had a post-operative cardiac arrest, from which she was resuscitated. There followed a severe delirium, with agitation, frightening visual hallucinations, disturbed nights, altered consciousness, and misrecognition of family members. There was no improvement on increasing doses of thioridazine – she started falling when getting out of bed to rescue people being attacked by

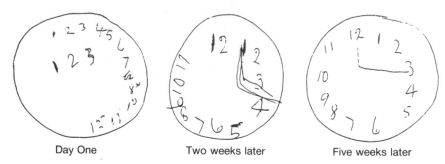

| Day One | Two weeks later | Five weeks later |

Fig. 2.2 Improvement in clock-drawing following recovery from toxic delirium (Shulman *et al.* 1986).

dinosaurs. Apart from a statement to the effect that she was 'aggressive', no comment was passed on this behaviour by the nursing staff until they were asked by relatives what investigations were being carried out: "She was confused before, wasn't she?" was their surprised response.

There is an extraordinary lack of interest by hospital doctors in elderly peoples' clinical and personal histories. Even after a prolonged stay in hospital, and frequent visits by concerned relatives and friends, the case-notes of an elderly patient with cognitive impairment and physical illness usually contain little more than terse statements such as 'collapsed' or 'confused' or 'poor historian', followed by a ritual review of the mental state usually recorded as an absence of symptoms. By contrast, the physical examination is often quite properly detailed. Despite this attention to physical detail, however, defeatist and ageist attitudes towards the elderly seem to raise the threshold of clinical suspicion, and potentially treatable conditions are not identified.

In view of the high prevalence of delirium among elderly medical and surgical in-patients, there may be a role for routine screening in these particular high risk groups. To date, however, instruments such as the MMSE that screen for disturbed cognition have not been shown to be effective in distinguishing delirium from other forms of cognitive impairment (Anthony *et al.* 1982). In order to do this, these instruments would need to be administered repeatedly, or else expanded to include items covering the history and course of cognitive impairment, as Millar (1981) attempted with the Clinical Interview Schedule (CIS). Serial clock-drawing appears to be a sensitive indicator of change in cognitive function (Fig. 2.2), but it is not yet known if errors on this test are diagnostically specific (Shulman *et al.* 1986). The Saskatoon Delirium Checklist (Miller *et al.* 1988) and the High-Sensitivity Cognitive Screen (HSCS) (Faust and Fogel 1989) may prove to be useful screening instruments for delirium. For now, however, cognitive screening for delirium with existing instruments is likely to be useful primarily as a consciousness-raising exercise for the medical and nursing staff involved.

An alternative approach to screening is to identify physical features that are predictive of delirium in elderly in-patients. Levkoff *et al.* (1988) examined the case-notes of 1285 elderly hospital admissions who received a discharge diagnosis of delirium, and identified four independent predictive factors:

(1) a urinary tract infection during in-patient stay;

(2) a low serum albumin on admission;

(3) a raised white blood cell count on admission; and

(4) proteinuria on admission.

These factors were also found to be predictive of delirium in a prospective study of 471 in-patients.

Differential diagnosis

The differential diagnosis of delirium in the elderly includes most other organic and functional psychiatric disorders in this age group, together with a few physical disorders. Many of these conditions are direct or indirect predisposing factors for delirium in the elderly, so it is important to bear in mind that their presence does not exclude the possibility that the subject is delirious as well.

Dementia

Distinguishing between delirium and dementia in an elderly patient is important primarily because of the grave consequences of mis-diagnosis if the subject is only suffering from a potentially reversible delirium. In practice, it is reliable information about the onset and course of the disorder from others who know the patient well that will establish the diagnosis. In contrast to delirium, dementia has an insidious onset and is likely to have been developing over months to years rather than hours to weeks. There are also important differences in the phenomenology of delirium and dementia. There is a global impairment of cognitive functions in dementia, but no impairment of wakefulness, and no diurnal fluctuation of cognitive impairment and alertness. The presence of hallucinations, mis-identifications and false beliefs are not diagnostically specific of delirium, since they occur quite commonly in uncomplicated dementia (Rabins and Folstein 1982). However, as Drinka *et al.* (1988) have pointed out, such phenomena should not be attributed to dementia unless intercurrent delirious states have been excluded. Physical signs such as tachycardia and pyrexia are more common in delirium than dementia (Rabins and Folstein 1982), and may be helpful in distinguishing between them. Where delirium is prolonged and subacute, it may come to resemble dementia more closely.

Demented individuals are at increased risk of developing delirium (Morse and Litin 1969; Hodkinson 1973), and it has been estimated that about one third of hospitalized demented patients are also delirious (Royal College of Physicians 1981). This predisposition is largely the result of a non-specific vulnerability to physical insult associated with brain damage (see Chapter 3), but in cerebrovascular dementia there may be a common aetiology, with patients presenting in delirium following a stroke (Balter *et al.* 1986). The clinical picture of delirium is altered somewhat in the context of dementia; in general, cognitive deficits are accentuated and hallucinatory and delusional phenomena are less prominent. In cases where the patient is known to be demented, any sudden deterioration in cognitive function and behaviour should be regarded as delirium until proved otherwise. In elderly patients who become disturbed following surgery, much of the difficulty in judging the degree of acute impairment and subsequent recovery could be avoided if their mental state was routinely assessed pre-operatively as a baseline (Lishman 1987).

Other organic mental disorders

Various other organic disorders may be mistaken for delirium, particularly if they have an acute onset. In transient global amnesia a brief episode of ischaemia in the distribution of the posterior cerebral circulation causes episodes of profound memory impairment that are of abrupt onset and last from hours to several days (Mathew and Meyer 1974). As in delirium there is disorientation in time and place and severe impairment of new learning ability, but there is no other cognitive impairment. In particular, the patient remains alert and there is no impairment of attention or perceptual disturbance. The abruptness of the onset also clearly distinguishes this disorder from delirium.

In the amnestic syndrome (Korsakoff's psychosis) there is persistent impairment of both short- and long-term memory. As a result, there is disorientation in time and place, and the patient may confabulate. There is no impairment of alertness.

Dysphasia due to a cerebrovascular accident in the dominant hemisphere is another focal cerebral dysfunction that has been mistaken for delirium on occasions (Grimley Evans 1982). Full cognitive function testing will demonstrate the localized nature of the problem, as will the absence of any attention deficit – unless, of course, the patient has been rendered delirious by the stroke.

Delusions and hallucinations secondary to specific organic factors may occur in the absence of disturbed alertness and severe cognitive impairment. In the organic delusional syndrome the delusions are typically persecutory in nature and are accompanied by psychomotor disturbance (both hyperactive and hypoactive) and dysphoric mood. Minor cognitive impairment may be present. In organic hallucinosis the hallucinations are persistent or recurrent, and may occur in any modality, most commonly visual and auditory. Delusions, if present, are not prominent in organic hallucinosis, and psychomotor disturbance is much less than that accompanying the organic delusional syndrome. These conditions occur in association with certain forms of epilepsy (temporal, occipital), following cerebral lesions (particularly in the non-dominent lobe), and may be induced by psychoactive substances. In the elderly, alcohol and prescribed drugs such as bromocriptine are more likely to be the cause of these states than substances such as LSD or amphetamines. Organic hallucinosis in old age may be due to sensory impairment, or to lesions in the peripheral sensory pathways as in the Charles Bonnet syndrome (see page 32).

Depression

The mental state in severe late life depression may at times resemble that of delirium, with agitation, retardation, persecutory auditory hallucinations and delusions, slowed thinking, impaired concentration, disturbed sleep and disorders of behaviour such as incontinence and screaming. Furthermore, any diurnal variation in the severity of depressive symptoms may be mistaken for the delirious fluctuation of cognitive impairment. It is sometimes very difficult to distinguish clinically between severe depression and delirium, but as a rule, if clinical and psychological testing show that the cognitive impairment is mild compared with the disturbances of mood, psychotic symptoms and psychomotor behaviour, it is likely that the patient is depressed rather than delirious. As with dementia a collateral history is helpful here; typically, the onset of depression is more insidious than that of delirium, and there may be a history of previous episodes of depression. The EEG in elderly subjects with depression is usually normal.

In old age, it is not uncommon for depression and delirium to occur together in the same patient. Late life depression is associated with physical illness and self-neglect, so depressed elderly people are at increased risk of developing delirium. Grief following bereavement may precipitate delirium in vulnerable individuals (see Chapter 3). The suicide rate associated with depression is increased in the elderly, and unsuccessful attempts by drug overdose or poisoning with carbon monoxide from car exhaust fumes may present as delirium to the accident and emergency services.

Mania

Episodes of mania are less common than depression in old age, but when they occur they can closely resemble hyperactive delirium. Subjects are distractable, irritable, agitated, disoriented, overactive, and they may have hallucinations and be deluded. The mood is usually labile and mixed rather than euphoric. In some cases there is a history of bipolar affective disorder but in others the mania is of late onset, possibly secondary to some cerebral insult (Shulman 1986). Mania secondary to levodopa and steroids has been described in the absence of delirium in younger patients (Krauthammer and Klerman 1978). Rigby et al. (1989) have reported mania in the absence of an organic brain syndrome following benzodiaze-pine withdrawal in an 83-year old woman, and suggest that the withdrawal state acted as a life event in precipitating the manic episode. In general, the first presentation of manic symptoms in late life is always an indication for investigation for underlying physical disorder, such as a stroke (Dunne et al. 1986).

There are several possible explanations for the mixed clinical picture of 'acute delirious mania' or 'manic delirium' seen in elderly patients (Bond 1980; Swartz et al. 1982; Shulman 1986). Elderly manic patients are particularly vulnerable to delirium. Untreated, they neglect themselves and rapidly become exhausted and dehydrated, and by the time they present it is likely that a proportion are suffering from both mania and delirium. Alternatively, the co-occurrence of mania and delirium may be the result of single underlying cause, such as tricyclic antidepressant medication in bipolar patients, or a cerebrovascular lesion in the non-dominant hemisphere (see Chapter 4).

Anxiety states

Panic attacks and panic disorder are rare in the elderly (Von Korff et al. 1985), but phobic disorders are among the commonest psychiatric dis-orders in the general elderly population (Regier et al. 1988; Lindesay et al. 1989). Elderly phobics have usually arranged their lives so that feared objects and situations are avoided, but circumstances occasionally make this impossible. For example, emergency admission to hospital will be very distressing for an individual with a blood/injury or needle phobia. Severe autonomic anxiety following exposure to a phobic stimulus may be mistaken for the fear associated with delirium. Other features of delirium are usually absent, but it has been suggested that catecholamines released peripherally during autonomic arousal may cross the elderly blood–brain barrier in sufficient quantities to precipitate a genuine delirium in some cases (Grimley Evans 1982). It has also been suggested that peripheral catecholamines may also be able to precipitate delirium by altering cerebral blood flow (Loach and Benedict 1980).

Paranoid states and schizophrenia

As a rule, these disorders are unlikely to be confused with delirium in the elderly. Although persecutory delusions and hallucinations may be prominent, particularly in paranoid states, the delusions are fixed and persistent and the hallucinations are usually auditory rather than visual. Schizophrenic oneroid states, in which patients become absorbed in dramatic and fantastic hallucinatory experiences, may resemble delirium in young adults, but they are extremely rare in old age. Elderly people with a long history of schizophrenia sometimes have attentional and other cognitive deficits, but in late-onset paranoid states there is usually no cognitive impairment.

A large proportion of the long-stay elderly population in mental hospitals have chronic schizophrenia, and these individuals are at increased risk of delirium because of their age, physical infirmity and the neuroleptic and anticholinergic drugs they consume. Unfortunately, low staffing levels and inadequate training of nursing staff mean that this delirium often goes unnoticed until the patient becomes severely ill or behaviourally disturbed. As the old asylums close and this population moves into small, better staffed residential units, this situation may improve.

Catatonia

Catatonic states occur in schizophrenia and in organic disorders, such as encephalitis, epilepsy, and certain metabolic disorders. They occur rarely in the elderly, and can closely resemble delirium, with psychomotor disturbance ranging from excitement to stupor, autonomic arousal, urinary incontinence, hallucinations and delusions, perseveration and disordered or incoherent speech. The presence of associated movement disorders such as rigidity, posturing, stereotypies, mannerisms and waxy flexibility is indicative of catatonia. Interviewing patients following administration of sodium amytal has been described as useful in distinguishing between catatonia and delirium (Ward *et al.* 1978); typically, catatonic patients become more accessible whereas delirious patients get worse.

The neuroleptic malignant syndrome (NMS) associated with acute administration or withdrawal of dopaminergic drugs is a rare but important differential diagnosis of delirium. Its features include alteration of consciousness, catatonic or extrapyramidal muscle rigidity and hyperpyrexia, and without treatment it is often fatal (Levenson 1985). NMS occurs most commonly in young adults, but elderly cases have been reported (Kellam 1987). The serum creatinine phosphokinase (CPK) is characteristically raised, and this is a useful discriminatory test if the diagnosis is suspected. However, it should be borne in mind that the serum CPK may be raised for many other reasons, such as following physical trauma, an intramuscular injection, or a myocardial infarction.

Malignant hyperthermia is another condition resembling NMS, with muscle rigidity, hyperpyrexia and raised serum CPK. It is an inherited abnormal muscle reaction to certain anaesthetic agents such as fluothane, and may present in elderly people undergoing surgery for the first time in their lives. Unlike NMS there is no central component to malignant hyperthermia, although there may be a delirium secondary to the hyperpyrexia.

Other functional disorders

Delirium-like states due to hysterical dissociation occasionally present in old age, usually in response to a severe life stress. The patient complains of confusion and there is severe disorientation on testing; disorientation in person, often with an inability to remember their own name, occurs more commonly than in organic delirium. Elderly people who are reluctantly dependent on others may deliberately become 'confused' and incontinent in order to influence events or aggravate their carers. As a rule, the EEG will be normal in these hysterical and factitious cases. However, it should be borne in mind that such cases are rare, and mild or bizarre disturbances of cognitive function and behaviour in elderly people should not be dismissed as 'neurotic' or 'hysterical' without thorough investigation. It is also worth remembering that individuals who are prone to hysterical dissociation may develop conversion symptoms when they are delirious (Lishman 1987).

3 The causes of delirium

The paciente is redde, if it com of blode – and Citrine – if it come of colera . . .
Parafrenesis cometh of byndynge togyders and company of other membres – as of
a postume of the stomake or of the mother, and whan these membres ben broughte
to their owne former state – than the braine torneth ayen to his owne good state
. . . But if the postume be in the substaunce of the braine – than is the frenesie
worst and moost grevous.

Bartholomeus Anglicus (1535)

Introduction

Aetiological factors in delirium are usually classified as either predisposing
or precipitating, with the predisposing factors increasing an individual's
susceptibility to a range of immediate causes. In most accounts of delirium
(e.g. Lipowski 1980*a*; Grimley Evans 1982), only organic factors such as
physical illness or intoxication are accorded the status of precipitants;
other factors such as age-related physical changes, psychiatric disorder,
bereavement, relocation and sensory impairment merely predispose.
Applied rigidly, however, this aetiological model causes problems, particularly in the elderly, since it is clearly apparent that delirious syndromes
can occur without an identifiable organic cause. Lipowski (1983) has
suggested that 5–20 per cent of cases of delirium in the elderly fall into his
category of 'pseudodelirium'. In the series reported by Sirois (1988) no
'probable aetiologic factors' were identified in 41 per cent of cases, and of
those cases where a physical cause was identified some were dubious, since
the delirium remitted before the biochemical abnormality did. In the
subjects studied by Koponen *et al.* (1989*b*), delirium was reportedly due
to 'severe psychological and environmental stress caused by a major life
change in a deeply demented patient' in 9 per cent of cases, and to
affective psychosis in 4 per cent.

The problem posed by this large residue of organically inexplicable
delirium in the elderly is not simply nosological; it represents a large group
of patients who are frequently referred to psychiatric liaison services
precisely because no sufficient organic aetiology can be discovered. If it is
allowed that the aetiology of delirium is often multi-factorial in the elderly

(Rabins and Folstein 1982), and that the factors involved are not only physical but also psychiatric, psychological, and environmental, then the diagnostic problem becomes one of assessing the relative contributions of these elements to the total picture rather than pursuit of the particular straw that happened to break the camel's back.

Physical factors

Drug therapy

Drug therapy is commonly implicated in the aetiology of delirium in the elderly. In Cutting's series (1980), it was considered the main factor in 16 per cent of cases and a contributory factor in a further 27 per cent. Almost any drug can precipitate delirium in an elderly individual; a list of those commonly involved is given in Table 3.1.

The increased sensitivity of the elderly to the effects of drugs is due primarily to age-related pharmacokinetic changes. Drug absorption is relatively unimpaired in the elderly (Stevenson *et al.* 1979), but there are significant alterations in the distribution, metabolism and excretion of drugs. Ageing is usually accompanied by a decrease in body weight, and elderly people require lower dosage regimens as a result. There is also an increase in the proportion of body fat with age, which will lower the plasma concentrations of lipid-soluble drugs such as tricyclic antidepressants, and increase the plasma concentrations of water-soluble drugs. The increased distribution volume of lipid-soluble drugs in the elderly also increases their elimination half-life (Braithwaite 1982). The plasma-protein binding of drugs does not appear to be consistently altered in the elderly; plasma albumin declines with age, but this may be compensated for by an increase in other plasma proteins (Cammarata *et al.* 1967).

The effect of ageing upon drug elimination depends upon the mechanism of clearance. Glomerular filtration rate is lower in the elderly, and this results in impaired excretion of water-soluble drugs. Most drugs are metabolized in the liver, and there is some evidence of impaired elimination by this route in the elderly as a result of reduced hepatic blood flow, liver mass and microsomal enzyme function (Stevenson *et al.* 1979). The clearance rate of drugs that are metabolized by oxidation, hydroxylation, and reduction (Phase I) is more consistently reduced in the elderly than that of drugs that are conjugated by glucuronide or sulphate (Phase II).

The changes in pharmacokinetics with age are inconsistent, and compounded by the influence of other factors such as sex, chronic disease, diet, smoking and the consumption of other drugs, including alcohol. As a result, there is a greatly increased variability in plasma concentrations of

Table 3.1 Drugs inducing delirium

Tranquillizers and hypnotics	Benzodiazepines Barbiturates Phenothiazines
Cardiovascular agents	Digoxin Diuretics Beta-blockers
Anticholinergic drugs	Atropine Hyoscine Antiparkinsonian drugs Tricyclic antidepressants
Dopamine agonists	Levodopa Bromocriptine Amantadine
Antibiotics and antituberculous drugs	Penicillin Streptomycin Sulphonamides Rifampicin Isoniazid Cycloserine Cephalosporins
Anticonvulsants	Phenytoin Sodium valproate Carbamazepine
Miscellaneous	Cimetidine Chloroquine Oral hypoglycaemic drugs Alcohol Analgesics

drugs following a given dose regimen in the elderly, with the effect that unpredictable toxic effects can occur at apparently therapeutic doses.

Certain drugs and drug groups are particularly associated with delirium in the elderly, perhaps because of age-related pharmacodynamic changes in the number or sensitivity of receptor sites. For example, any agent with anticholinergic activity may precipitate delirium in the elderly (Lipowski 1983), and this may be related to the depletion in central cholinergic systems that occurs both as part of the normal ageing process and degenerative cerebral disease (Blazer *et al.* 1983; Sunderland *et al.* 1987) (see Chapter 4).

Thus, drugs such as tricyclic antidepressants, procyclidine and benz-hexol, which the patient may have been taking for many years for depression or Parkinson's disease, may contribute to the development of delirium in the presence of some additional insult. Reduction of the dosage or complete discontinuation may prove extremely worthwhile.

Another problematic group of drugs are the dopamine agonists, such as levodopa and bromocriptine. There is a particular tendency for this group to produce hallucinations in the elderly, thus making the treatment of Parkinson's disease very difficult in some cases.

The causes of drug-induced delirium can sometimes be quite surprising. Penicillin is frequently regarded as one of the safest of antibiotics, if allergic reactions are excluded. Nevertheless, large doses over prolonged periods in patients with impaired renal function can result in neurotoxic effects such as convulsions and delirium, for example during the treatment of infective endocarditis (Mandell and Sande 1985).

Infection

Drug therapy may be the commonest cause of delirium in the elderly, but provided that it is considered in the first place it is not a diagnostic problem, since the patient's medication is usually known to the attending doctor. By contrast, infection is not always easily diagnosed in the elderly, and delirium may be the most prominent presenting feature. The pyrexial response to infection is usually present in the elderly, but it may be blunted and delayed (Downton *et al.* 1987) and so appear to be absent (Berman and Fox 1985). The white cell count may rise less than in younger adults, and in some severe pneumonias it may even fall as neutrophils are taken up into the infected area of lung. The erythrocyte sedimentation rate (ESR) is not a particularly useful indicator of infection in the elderly, since it may be elevated from a variety of other causes, such as rheumatoid disease or polymyalgia rheumatica. To complicate matters further, these indicators may be abnormal for reasons other than infection: for example, painless myocardial infarction may cause delirium with pyrexia, a high white cell count and a raised ESR. The diagnosis of infection may be obtained by measuring C-reactive protein (CRP) levels (Rose *et al.* 1981), but in most hospitals there is too long a delay before the result is available for this to be diagnostically useful, and its greatest value is in the monitoring of the efficacy of long term treatment. It is more likely, therefore, that the presence of infection will be identified with greatest certainty by establishing its site.

Urinary tract infection

Elderly patients tend not to report the typical symptoms of a urinary tract infection (UTI). Incontinence of urine is common in the absence of infection, and the onset of falls and the worsening of mental state become more common as presenting symptoms of a UTI than dysuria and urinary frequency as patients get older. Simple dipstick testing for the presence of blood and protein may suffice to confirm the diagnosis, but it should be remembered that on occasions these tests yield false negative results.

Patients with indwelling catheters, calculi, or prolapsed uterus are at increased risk of delirium, and grumbling urinary infections often present as episodic deterioration of their mental state. These episodes are quite frequently associated with the finding of bacteraemia on blood culture. In such cases, long-term antibiotic prophylaxis should be considered seriously. Trimethoprim is safe and often effective but if patients have multi-resistant Gram-negative organisms, then ciprofloxacin may be tried, although formal trials of such a practice have not yet been carried out.

Pneumonia

Pneumonia is one of the classical causes of delirium in younger adults, and it is also a common cause in old age. The frequency of pneumonia as a reported cause of delirium in the elderly is close to that of urinary tract infection (15 per cent against 17.5 per cent respectively in one study (Hodkinson 1973)). Both disorders are associated with septicaemia and bacteraemia, but it is not known whether there is an association between septicaemia or bacteraemia and delirium. However, bacteraemia is one of the features associated with poor prognosis in pneumonia, and it is a useful diagnostic indicator of the causative organism.

On occasion the patient presents with delirium and a mild pyrexia only, and it is 24–48 hours before physical signs develop in the chest, or the chest radiograph shows consolidation. Most episodes of pneumonia that occur outside hospital are caused by pneumococci, but so-called 'atypical' pneumonias, due to organisms such as *Mycoplasma pneumoniae* and *Legionella*, are becoming increasingly common. One of the prominent clinical features of atypical pneumonia is the degree of mental disturbance, particularly in *Legionella* infection, but prospective studies have not shown that this feature is helpful diagnostically, since mental disturbance may occur with all forms of pneumonia (Woodhead and MacFarlane 1987). Reactivation of tuberculosis must not be forgotten as a possible cause of pneumonia in an elderly individual.

The precise mechanism of delirium in pneumonia is unclear, and it is likely that a number of features may contribute, each of which may cause delirium in its own right. Thus hypoxia, hypercapnia, pyrexia, septicaemia, hypotension ('shock'), reduced cerebral blood flow, and dehydration may all play a part.

Septicaemia

As has been mentioned, this condition is clearly associated with delirium, but it does not usually occur as an isolated event. As a rule, the bacterial toxins that cause septicaemia arise from a focus of infection that should be searched for, found, and treated. Bacteraemia is defined as the presence of a viable organism in the blood, diagnosed by blood culture. If bacteraemia is persistent it also implies the presence, and need for identification, of a source of infection. The phenomenon of transient bacteraemia following dental work or urinary tract manipulation is well recognized, but truly transient bacteraemia does not cause delirium.

Infective endocarditis

Infective endocarditis is often difficult to diagnose, and this is especially true in the elderly. The development of delirium following one of the procedures causing bacteraemia mentioned above may be an early sign that a persistent focus of infection has become established. A history of rheumatic or valvular heart disease is frequently lacking in the elderly, and the presence of mitral valve prolapse caused by ischaemic heart disease may provide sufficient abnormality for the establishment of infection. Other features on presentation may give a clue as to the site of infection, for example the development of left-sided heart failure. Diagnosis is best achieved by maintaining a high index of suspicion, by carrying out repeated blood cultures, and by performing echocardiography to identify any vegetations that may be present (Cantrell and Yoshikawa 1984). The CRP level may be very high, and if so it will provide a useful guide to the efficiency of treatment, since antibiotic therapy has to be maintained for several weeks.

Ulcers, pressure sores and gangrene

Superficial ulcers or pressure sores are unlikely to give rise to delirium, but when they become deep and infected the patient may well become delirious. It is likely that organisms colonizing these deep and necrotic wounds are a cause of septicaemia, but firm evidence is lacking. Such cases require treatment with systemic antibiotic therapy once the organism has been identified from the wound or from blood cultures.

So long as gangrene remains uninfected ('dry gangrene') there is little systemic toxicity, but with the onset of infection the situation changes rapidly and the level of mental deterioration then parallels the extent of the necrosis. With inadequate blood supply, the local defence mechanisms in gangrenous tissues are poor, and infection commonly results in generalized septicaemia. The improvement in mental state that follows amputation of an infected ischaemic foot can be dramatic.

Intracranial infections

Meningitis and encephalitis are unusual infections in the elderly but frequently cause delirium when they occur, and must be borne in mind in the differential diagnosis. Tuberculous, bacterial or viral meningitis may present as delirium unaccompanied by the classical signs of meningism, and less typical organisms such as *Listeria monocytogenes* are becoming more common as causes of meningitis in the elderly (Rau and Voegt 1972). Encephalitis is usually viral, and it is important to determine whether or not it is due to herpes simplex, since infection with this organism can be treated with acycloguanosine (Acyclovir). Early treatment is imperative, since subsequent cerebral damage is proportional to the delay in starting treatment. Finally, it should not be forgotten that syphilis may present as delirium in old age.

CSF examination should be undertaken if specific neurological findings are absent and no other site of infection has been identified. The rapid onset of delirium accompanied by changing neurological signs may be due to a space-occupying cerebral abcess, and urgent CT scanning is indicated.

Metabolic, endocrine, and nutritional causes

Electrolyte abnormalities

Delirium may be produced by dehydration or as a result of a fall in cerebral perfusion, and may be exacerbated by the cause of the dehydration, for example uncontrolled diabetes mellitus or hypercalcaemia. Rapid rehydration with 5 per cent dextrose when a patient has been hyperosmolar for some days may result in cerebral oedema as osmotic equilibrium is established across the blood–brain barrier. However, when the serum sodium is high it is unwise to give normal saline, since renal function and the ability to excrete excess sodium are likely to be impaired. A process of controlled rehydration should be adopted instead.

Delirium caused by water overload may result from inappropriate secretion of antidiuretic hormone (Edwards 1977), which may occur in patients with cancer, with pneumonia, and with intracerebral lesions including haemorrhage, infarction and tuberculous meningitis. As serum sodium levels fall below 115 mmol/litre there is an increasing probability of epileptic fits developing, thus further complicating the picture.

A fall in phosphate levels is an important but sometimes neglected cause of delirium. It may occur during prolonged nutritional support, or in hyperparathyroidism. In hyperparathyroidism, delirium is most likely if calcium levels are high (Kleinfield *et al.* 1984).

Diabetes mellitus

In patients with diabetes mellitus, hyperglycaemia and ketoacidosis may give rise to delirium at any age. In the elderly, hyperosmolar non-ketotic coma is especially common and frequently presents as delirium. Furthermore, the elderly are at particular risk of delirium in association with hypoglycaemia. Firstly, hypoglycaemia becomes more common if relatively long-acting oral antidiabetic agents such as glibenclamide are used, since their half-life is increased in the presence of reduced renal and hepatic function. The elderly do not produce the same autonomic response to hypoglycaemia as do the young, and it seems that normal cerebral function takes longer to return after the blood glucose level has been normalized. Thus patients may be diagnosed erroneously as having dementia when in fact they are becoming hypoglycaemic in the early hours of the morning, and fail to regain normal cerebral function before their next episode of hypoglycaemia. Delirium may be the presenting feature of insulinoma in the elderly (Service *et al.* 1976).

Renal and hepatic failure

Uraemia, hyperphosphataemia, and acidosis contribute to the cerebral effects of renal failure. Liver failure, whether acute or chronic, may lead to hepatic encephalopathy, which progresses rapidly through delirium to coma if left untreated. Both renal and hepatic failure may cause delirium by reducing the clearance of drugs.

Endocrine disorders

Hypopituitarism may give rise to delirium, perhaps as a result of water retention secondary to lack of cortisol; however, delirium is uncommon in adrenal insufficiency unless caused by hyponatraemia. 'Myxoedema madness' still occurs, but given the usual slow development of the primary disorder, mental impairment in hypothyroidism is more likely to be chronic rather than acute. In thyrotoxicosis delirium may be confused with, or combined with, manifestations of sympathetic over-activity such as agitation and anxiety.

Protein-calorie malnutrition

Hypoproteinaemia leads to increased transudation of fluid from the circulation into all body tissues including the brain, and in severe malnutrition hypophosphataemia, vitamin and trace element deficiencies also contribute to the mental impairment. When refeeding is begun it may be

necessary to give concentrated albumin to maintain correct water distribution until the liver is functioning sufficiently to produce albumin
endogenously.

Vitamin deficiencies

Wernicke's encephalopathy secondary to thiamine deficiency usually presents abruptly, with the characteristic triad of delirium, ataxia, and
ophthalmoplegia (Victor *et al.* 1971). It is associated with chronic alcoholism coupled with an inadequate diet, and rarely with other conditions
causing dietary deficiency, such as oesophageal stricture and gastric
carcinoma. Delirium may also occur with deficiencies of vitamin B_{12}
(Hector and Burton 1988), folic acid (Royal College of Physicians 1981)
and nicotinic acid (Spivak and Jackson 1977). In practice, vitamin
deficiencies caused by malnutrition will be multiple, and this should be
borne in mind when prescribing supplements.

Hypothermia

This condition may complicate many of the metabolic and endocrine
problems outlined above, and also a number of other possible precipitants
of delirium, such as myocardial infarction and stroke. Hypothermia alone
is believed by some to cause delirium (Maclean and Emslie-Smith 1977),
although others maintain that hypothermia does not occur in an otherwise
healthy individual of whatever age, and that it is the associated condition
that precipitates the delirium (Whittle and Bates 1979).

Cardiovascular and respiratory disorders

The disorders discussed in this section have as their common denominator
the failure of delivery of an adequate supply of oxygen to maintain normal
cerebral metabolism. As well as precipitating delirium, these disorders
also reduce the threshold for the development of delirium, either as a
result of diffuse cerebrovascular disease, or accelerated neuronal death
associated with chronic under-supply of oxygen.

Cardiac failure

Cardiac failure from all causes accounted for 24 per cent of cases of
delirium reported by Hodkinson (1973). A sudden fall in cardiac output
may be caused by myocardial infarction or by the onset of an arrhythmia
(Bergmann and Eastham 1974). A large pulmonary embolus will cause an
acute fall in cardiac output and hypoxia will exacerbate the effects on the
brain in this condition and in the presence of pulmonary oedema. Less
commonly, a fall in cardiac output results from pericardial tamponade
(infective, or following myocardial infarction), or constrictive pericarditis

(usually old tuberculosis or infiltration with tumour). The presence of ischaemic heart disease *per se* does not seem to predispose to delirium (Hodkinson 1973). Since myocardial infarction is often silent in the elderly, the sudden onset of delirium may be the presenting feature, and an electrocardiogram should be included in the routine investigations.

Vascular disease

Atheroma develops in coronary, peripheral, renal, and cerebral vessels. Therefore, ischaemic heart disease may be a marker for those patients at risk of a greater than normal fall in cerebral blood flow if hypotension develops as a result of septic shock, hypovolaemic shock, drug therapy with vasodilators of any type, or primary cardiac disease.

Elderly patients with hypertension of many years standing may be difficult to manage, since the normal response of the physician is to give treatment to reduce the blood presssure, and if this is done too rapidly an acute fall in cerebral perfusion may result. Hypertensive encephalopathy is a cause of delirium in those with severe or accelerated hypertension, but it is much less common in the old than in the young since in the old the brain is 'protected' from a high systemic blood pressure by proximal arteries narrowed by atheroma.

Collagen vascular disorders such as systemic lupus erythematosus may affect the cerebral vasculature diffusely, but this is a rare disorder, particularly in the elderly. The presence of systemic manifestations, a preceding history, and a high ESR may provide a clue to the diagnosis. Cranial arteritis, alone or in association with polymyalgia rheumatica, is more common in the elderly, but usually presents as a rare cause of stroke rather than as delirium.

Other respiratory disorders

Acute pulmonary disorders other than pneumonia and pulmonary embolism that may cause sudden hypoxia and thus delirium include pneumothorax and very large pleural effusions. The chief role of chronic lung disease in the aetiology of delirium is that patients have less reserve of oxygenation and sometimes carbon dioxide excretion, so that other disorders such as pulmonary infection or a fall in cardiac output are more likely to lead to cerebral hypoxia than in individuals with normal lungs. However, many patients with severe chronic lung disease become 'acclimatized' to hypoxia at levels which would be almost certain to produce mental impairment and possibly delirium were they to be produced suddenly in a normal elderly person, for example an arterial PO_2 of 5–6 kPa. Some patients with chronic lung disease are prone to episodes of anxiety with hyperventilation; if this is severe or prolonged, the resulting alkalosis may precipitate delirium.

Individuals who are noticeably more cognitively impaired on first waking in the morning may be suffering from CO_2 retention during sleep at night, or from hypoxia due to sleep apnoea (Kales *et al.* 1985). These phenomena have not yet been fully investigated in the elderly (Berry *et al.* 1987).

Anaemia and polycythaemia

Whatever the cause of the anaemia, the result is a reduction in the oxygen carrying capacity of the blood. If this cannot be matched by a compensatory increase in cardiac output, then tissue hypoxia will result. In narrowed vessels the fall in blood viscosity that results from a drop in haemoglobin level down to 12.5 g/dl allows a concomitant increase in flow-rate, so that transfusion to strictly 'normal' levels is not necessary.

Conversely, blood viscosity increases exponentially with increasing haematocrit, but there is only a linear increase in oxygen carrying capacity. Thus cerebral oxygenation will be reduced progressively as the haematocrit rises above 50 per cent (haemoglobin about 16 g/dl). This may occur in chronic hypoxia caused by chronic lung disease, in very heavy smokers where much of the haemoglobin is occupied by carbon monoxide, or in polycythaemia rubra vera. Rarely, chronic myeloid leukaemia may cause a significant increase in blood viscosity with the same effects on cerebral oxygenation.

Intracranial lesions

Trauma

Following a severe blow to the head there is often a period of memory loss which may be retrograde, i.e. concussion. This is often accompanied by disorientation and reduced alertness in the elderly, but should recover steadily. Severe head injury may cause cerebral contusion, often situated on the opposite side of the brain to that of the initial injury (the 'contre-coup' lesion). Contusion is suggested by lateralizing neurological signs which accompany the delirium, and should also recover fully and steadily. The development of a subdural haematoma is characterized by progressive deterioration or fluctuation in conscious level over a few days, the lateralizing neurological signs being less prominent in relation to the degree of mental impairment than is the case in stroke. Previously undiagnosed subdural haematomata are a fairly common finding at autopsy in the elderly, and since the success of neurosurgical evacuation decreases with advancing age, asymptomatic chronic subdurals are best left alone.

Stroke and subarachnoid haemorrhage

Delirium has been reported as the presenting feature in 13 per cent of major strokes (Dunne *et al.* 1986), and as a frequent presenting syndrome in small localized infarctions (Medina *et al.* 1974; Balter *et al.* 1986). Indeed, delirium is well-recognized in association with infarction in the territory of the right (non-dominant) cerebral artery, and may be the only abnormality observed (Mesulam *et al.* 1976; Mori and Yamadori 1987). Delirium also occurs in posterior cerebral artery infarcts (Devinsky *et al.* 1988).

The presence or absence of delirium cannot be used as a feature to differentiate haemorrhage from infarction, but since unconsciousness at onset is a feature of haemorrhage it may be expected less often in the early stages of that condition. Delirium has also been reported in association with subarachnoid haemorrhage.

Secondary effects of infarction such as localized cerebral oedema may give rise to a period of delirium coming on 24 hours after the initial event, but other causes such as superimposed chest infection must also be considered.

Tumour

Primary, or more commonly secondary, cerebral malignancy may present in the elderly as a stroke with subsequent downhill progression, with episodes of partial recovery and then further progression, or with intermittent or progressive delirium either alone or in combination with localizing neurological signs. Delirium is also a non-metastatic complication of certain tumours, notably bronchial carcinoma, and may be the first clinical manifestation of the malignancy. Carcinomatosis accounted for 5 per cent of cases of delirium reported by Hodkinson (1973).

Epilepsy

In the elderly, post-ictal confusion may be more prolonged than in younger individuals, and if the patient is having numerous minor or incomplete epileptic fits the next ictus may occur before the delirious phase has resolved. A bitten tongue, cheek or lip is strongly suggestive of epilepsy in a delirious patient. If the patient is a known epileptic, it is important to measure the blood levels of anticonvulsants, particularly phenytoin, since delirium is also a major feature of phenytoin toxicity (Schmidt, 1982).

The rare phenomenon of epilepsia partialis continua may manifest itself as delirium, but as its name suggests it is in fact a form of minor status epilepticus. Diagnosis is confirmed by EEG.

Intra-abdominal causes

Inflammatory conditions

The mechanism of induction of delirium in this group of conditions is probably a combination of hypotension due to fluid loss from the circulation, and infection or chemical inflammation. Perforation of a gastric ulcer or of a diverticulum is usually followed by peritonitis, but this may be relatively 'silent' in the elderly. The patient loses large quantities óf fluid into the inactive gut and inflamed peritoneum. Local defence mechanisms are poor, and surgery for the perforated viscus is usually necessary. Acute pancreatitis typically presents with acute abdominal pain and shock, or less acutely with hypotension, pyrexia and ill-defined abdominal tenderness. If diagnosed and treated early, acute pancreatitis has a surprisingly good prognosis in many patients of this age group. The assessment of patients with delirium and any hint of intra-abdominal pathology should include an erect abdominal X-ray and a serum amylase level.

In the biliary tract, gall-stones may not cause severe pain in an elderly person, but may give rise instead to bouts of pyrexia and delirium, with nausea and vomiting. The infectious element may predominate in the form of ascending cholangitis. This diagnosis is revealed initially by observing fluctuation in hepatic enzyme levels, and then by ultrasonic identification of calculi, sometimes found in the common bile duct after cholecystectomy.

Non-inflammatory conditions

Upper gastrointestinal haemorrhage may present as delirium before hypotension becomes manifest. The cool, shut-down periphery and active bowel sounds, without evidence of left-sided heart failure, often provide clues, and the true situation is usually revealed quite soon with the first haematemesis or melaena stool.

Constipation is a well-documented cause of delirium in the elderly, and may be so severe that the bowel is effectively obstructed. Relief of the blockage is a matter of some urgency since it may be followed by inflammation and subsequent perforation of a colonic diverticulum.

Alcohol and drug abuse

Alcohol abuse does not appear to be a major problem in the present elderly population in the UK. In a random community sample of people aged 65 years and older in Liverpool (Saunders *et al.* 1989) less than one per cent were identified as having current drinking problems. However, the rate may be higher in other countries, and it is likely to increase in all developed societies as more affluent cohorts of men and women accustomed to higher levels of alcohol consumption enter old age. The rate of alcohol abuse in elderly medical and surgical patients will be higher than that in the general population, and alcohol withdrawal should always be considered as a possible cause of delirium in this group, particularly if it develops for no apparent reason about 48 hours after an emergency admission. Elderly alcoholics who are dependent on a State pension are

Table 3.2   The CAGE questionnaire (Mayfield *et al.* 1974)

---

1.   Have you ever felt that you ought to cut down on your drinking?

2.   Have people annoyed you by criticizing your drinking?

3.   Have you ever felt bad or guilty about your drinking?

4.   Have you ever had a drink first thing in the morning to steady your nerves or get rid of a hangover?

A score or two or more identifies subjects who may be abusing alcohol.

likely to be going without adequate food and warmth in order to fund their addiction, and will therefore be at greater risk of physical illness and delirium.

Doctors are sometimes rather shy about asking elderly people about their drinking habits, but an adequate alcohol history is essential if withdrawal states during hospital admission are to be anticipated. The four-item CAGE questionnaire (Mayfield *et al.* 1974) is a quick, acceptable method of identifying those who are likely to be abusing alcohol (Table 3.2). A longer, but more accurate screen is the MAST questionnaire (Selzer 1971). Elderly men are usually quite frank about their alcohol consumption, but women tend to conceal their drinking, and closer questioning and investigation may be required.

Alcohol is not the only drug of abuse and addiction in old age. Elderly people are heavy consumers of psychotropic medication: in general practice studies, the elderly are the heaviest users of psychotropic drugs, particularly benzodiazepine hypnotics and anxiolytics (Skegg *et al.* 1977; Catalan *et al.* 1988). About 10 per cent of patients over 65 years taking benzodiazepines become delirious following sudden withdrawal (Foy *et al.* 1986), and up to 45 per cent may develop a withdrawal syndrome including palpitations, tremor, depersonalization and derealization which may approach frank delirium in severity (Liappas *et al.* 1987). In such cases, features of the original anxiety state overlap broadly with many features of delirium, and differentiation is difficult, but important (Fontaine *et al.* 1984).

Withdrawal of shorter-acting benzodiazapines tends to produce a disturbance of the sleep pattern. Withdrawal symptoms tend to be most marked between the fifth and sixth day after admission/withdrawal (Liapass *et al.* 1987). The occurrence of epileptic fits suggests a withdrawal syndrome, and is associated with the use of about 100 mg per day of diazepam or equivalent (Preskorn and Denner 1977).

Opiates

The classical opiate withdrawal syndrome is a combination of autonomic symptoms and signs, muscle and abdominal cramps, and psychological symptoms such as anxiety and irritability. These symptoms usually peak two to three days after stopping the drug (Liappas *et al.* 1987), although for methadone withdrawal the interval is over twice as long (Hodding *et al.* 1980). Elderly addicts are rarely abusers; morphine and its derivatives will almost certainly have been prescribed and not obtained illegally. The experience of severe pain following the withdrawal of opiates in the elderly is a more likely cause of delirium than a withdrawal syndrome.

Less addictive agents such as pentazocine, buprenorphine, codeine, and dextropropoxyphene may on occasion produce withdrawal syndromes, although not usually amounting to fully developed delirium. All of these agents cause constipation, and may also cause hallucinations in certain individuals. Careful history taking may be required to establish the time of onset of the delirium in relation to starting or stopping the suspect agent.

Barbiturates

Elderly patients may have been taking barbiturate drugs for years because of idiopathic epilepsy, or may have become dependent on barbiturate hypnotics that were prescribed before their dangers were fully appreciated or alternatives were available. Barbiturate overdose is still a significant means of suicide in the elderly (Lindesay 1986), and it is important that they should be weaned off these drugs whenever possible. However, this should be done carefully, especially if large doses are being taken, since delirium, usually accompanied by epileptic fits, may be precipitated (Liappas *et al.* 1987). Hyperpyrexia is a characteristic accompanying feature of barbiturate withdrawal, and may prove fatal (Khantzian and McKenna 1979)

Surgery and anaesthesia

The factors that predispose the elderly to post-operative delirium provide a convenient summary of those conditions, discussed above, that can produce delirium on their own. These include abnormal electrolyte levels, elevated blood urea and creatinine, reduced albumin, underlying cardio-vascular disorder, peri-operative hypotension, pre-operative respiratory disease, infections including wound infection, intravenous infusions, urinary catheterization, opiate analgesia, and pre-operative organic or func-tional psychiatric disorder (Seymour and Vaz 1987; Seymour and Pringle

1983; Heller *et al.* 1979; Kornfeld *et al.* 1974; Berggren *et al.* 1987). Pre-medication with anticholinergic drugs has been associated with post-operative delirium following open heart surgery (Tune *et al.* 1981), hip fracture (Berggren *et al.* 1987) and cataract extraction (Van Deuren and Missotten 1979). Recently, a reduction in the incidence of post-operative confusion has been reported with the use of glycopyrrolate as an anticholinergic premedication, since this does not cross the blood–brain barrier (Sheref 1985).

Rarer causes of delirium

A number of unusual causes of delirium have been reported, including solvent-sniffing (Ron 1986) – probably inadvertent in the elderly, carbon monoxide poisoning (Smith and Brandon 1970), cerebral malaria (Toro and Roman 1978), sickle-cell disease, heatstroke, radiation exposure, electrocution, porphyria, and ingestion of poisonous plants or fungi (Davison 1989). The probable mechanisms of these insults will be apparent from the preceding discussion of more common causes.

Psychiatric disorders

Dementia

It is a common clinical observation that chronic organic disorders such as dementia predispose to delirium in the elderly, and one that is supported by most studies that have examined this issue (Hodkinson 1973; Bergmann and Eastham 1974). In the sample described by Koponen *et al.* (1989*b*), only 13 per cent had no apparent central nervous system disease. This predisposition is usually ascribed to the increased vulnerability of a damaged brain to acute physical insults. Evidence in support of this view comes from the CT scan studies by Koponen *et al.* (1987; 1989*a*), which have demonstrated a greater frequency of radiological abnomalities suggestive of dementia and Parkinsonian changes associated with delirium. However, Blessed *et al.* (1968) found a marked distinction between delirium and functional psychiatric disorders on the one hand and dementia on the other, in terms of the numbers of senile plaques in the brains of their 60 elderly in-patients who had been cognitively assessed and diagnosed in life. It should be borne in mind that the effect of physical illness upon cognitive function in elderly people may vary for reasons that are independent of any underlying brain disease. For example, factors such as altered immune response or impaired respiratory reserve can greatly increase the impact of apparently trivial disorders on cognition.

Depression and anxiety

Delirium has also been reported in association with depression by several investigators. In a retrospective series of 472 patients assessed following admission to hospital in the 1930s and 1940s, Roth (1955) described transient disorientation or 'clouding' in 12 per cent of the cases of acute affective psychosis. The physical condition of these patients was not reported, but the definition of 'acute confusion' did not require the presence of an organic cause. In their small sample, Bergmann and Eastham (1974) found that their delirious patients fell into two groups; those with underlying cerebral disease, and those with no physical cerebral disease but with long-standing psychiatric vulnerability, indicated by previous episodes of psychiatric illness, family history of anxiety and depression, and 'sensitive' and 'obsessional' personality traits. In suggesting that psychiatric treatment of anxiety and depression might prevent future delirium, they breached the traditional view that delirium was inextricably linked with organic cerebral pathology.

In Hodkinson's study (1973), depression was the only non-organic diagnosis associated with 'toxic confusional states', occurring in 12 per cent of the total sample and in 14 per cent of those whose mental test scores changed significantly. However, he dismissed any direct aetiological role for depression, attributing this association to poor test scoring because of depression, tricyclic antidepressants, or the association between depression and severity of physical illness, the latter causing the delirium. Similar points have been made by Lipowski (1983).

There is disagreement about the relationship between pre-operative depression and anxiety, and post-operative delirium. Millar (1981) found no relationship between delirium ('post-operative intellectual impairment') and pre-operative depression, past psychiatric history or family psychiatric history in his study. Similarly, Simpson and Kellet (1987) found no relationship between pre-operative anxiety and post-operative delirium in 45 patients examined before and after total hip replacement. Berggren et al. (1987) however, noted depression as an important predictor of post-operative delirium in their sample of elderly patients with fractured neck of femur. Sirois (1988) has criticized the tendency to explain anxiety symptoms as secondary to the underlying cognitive impairment, and to discount anxiety states as an aetiological factor. In his series of referred cases of delirium, he found that anxiety was more common in delirious surgical patients post-operatively than in delirious medical patients, and that in the surgical patients 'very often' no particular physical abnormality could be found to account for the state. Although it is possible that raised peripheral catecholamines in anxiety may precipitate delirium by various mechanisms (see Chapter 2), it should be borne in mind that the absence of an apparent 'organic' cause is no guarantee that none is operating. For example, the long-term effects of anaesthetics on cerebral function is still not clear (Blundell 1967; Drummond 1975).

An unusual relationship between delirium and depression has been suggested by Borchardt and Popkin (1987); they report two cases of chronic depression that improved following delirium caused by a severe physical illness. However, it is unclear to what extent the improvement in the depression was due to the delirium, or to the physical and psychological impact of the illness.

In summary, the part played by affective disorders in the aetiology of delirium is still not clear. However, if attention to affective disturbance might prevent delirium, as originally suggested by Bergmann and Eastham (1974), then surely this is a topic that deserves further study.

Schizophrenia

Although use of a term translated as 'delirium' is common in Eastern European and Soviet descriptions of schizophrenic phenomenology (e.g. Alimkhanov 1988), there is no equivalent tradition in Anglo-American psychiatry. There have been no studies to date of the association between delirium and schizophrenia in the elderly; the careful comparison of the phenomenology of the two conditions by Cutting (1987) used selected samples with one or other of the two disorders rather than a study population where they might have co-existed. Certain physical illnesses such as vitamin B_{12} deficiency rarely cause both delirious and schizophreniform states in elderly subjects (Hector and Burton 1988).

Factors associated with psychiatric disorder

Self-neglect is an important factor predisposing to delirium in the elderly, and is often the result of a psychiatric disorder. In particular, demented elderly people who live alone are at risk of becoming dehydrated, malnourished, vitamin deficient, hypothermic, and infected. Other psychiatric disorders which may result in severe self-neglect include depression, mania, chronic schizophrenia and paraphrenia. Elderly people with persistent psychopathic and inadequate personality disorders may also neglect and abuse themselves (Grimley Evans 1982). Sometimes personality disorders in old age result in reclusiveness, squalor, self-neglect and refusal of all services, the so-called 'Diogenes syndrome' (Macmillan and Shaw 1966; Clark et al. 1975). They often only present in extremis, and in one study of such people in the UK who had been compulsorily removed from home under Section 47 of the National Assistance Act, 24 per cent had a primary diagnosis of delirium at the time of removal (Wolfson et al. 1990).

The effects of psychiatric treatment

Psychiatric patients are also at increased risk of developing delirium because of the treatments they receive; in particular, psychotropic drugs such as tricyclic antidepressants and benzodiazepines are potent precipitants of delirium in the elderly. Electroconvulsive therapy (ECT) usually results in a short period of post-treatment cognitive impairment (Leechuy *et al.* 1988), and this can sometimes be prolonged and troublesome in the elderly. For this reason, it is usually recommended that it be given unilaterally to the non-dominant hemisphere only, since this shortens the recovery time (Fraser and Glass 1978). There is no evidence that unilateral ECT is less effective than bilateral in the treatment of depression (Fraser and Glass 1980). In some cases, delirium associated with ECT may be due to the anticholinergic premedication.

Personality

Personality assessment is a popular activity, and it is not surprising that personality traits have been studied as possible predisposing factors for delirium. However, assertions from clinical experience that certain personality types such as the 'anxiety-prone' or 'dominant' are more liable to delirium than others (Grimley Evans 1982) are bound to be misleading, since any clinician confronted with a patient in the throes of a delirium is hardly in the best position to assess the person's pre-morbid personality. The assessment of personality in the symptom-free is problematic enough, and psychiatrists cannot make valid comparisons from clinical practice since they are not called to an equal number of non-delirious patients by their physician colleagues. Consider the attempt by Scott (1960) to relate pre-operative personality to 'post-operative psychosis'. He could only find one patient in his series who 'did not exhibit any unusual personality traits'. Two were 'oversensitive, apprehensive, unstable, suspicious, and they responded to stress with undue anxiety'. The remaining eight were 'presenile'; that is 'stubborn, unco-operative, querulous or hostile' under stress. Scott contrasted these patients with 'the average patient', who 'easily absorbed' the stresses of admission.

Another problem with studies implicating personality traits as predisposing factors for delirium is that they rely for the most part on retrospective judgements. For example, in the study reported by Morse and Litin (1969) 60 post-operative patients over 30 years old were referred over six months by nursing and medical staff to the study on account of perceived disorientation. The investigators defined delirium as an acute syndrome of disorientation, memory, judgement and intellectual impairment, and lability of affect: they did not mention clouding of consciousness or fluctuation of impairment. There were 57 post-operative controls, matched for age, sex, and type of operation. As far as one can tell, no pre-operative assessments were carried out. The psychological factors found to be more common in the delirious group were pre-operative fear of death, failure to deny pre-operative anxiety, a 'clinical diagnosis of paranoid personality' and retirement-adjustment problems. In addition, 37 per cent of the delirious group and none of the controls were reported as having an underlying organic brain syndrome, which itself may account for many of the differences observed.

The methodological problem with retrospective studies is obvious: how can one disentangle the effects of delirium from its causes, when one only sees the patient after the event? Some studies of delirium following surgical procedures have made use of pre-operative assessments of personality. Blundell (1967) found no significant contribution of personality to delirium in her prospective study of elderly surgical patients. Millar (1981) compared surgical patients who developed delirium with those who remained cognitively intact, and found no association between delirium and pre-operative psychiatric status or environmental factors. In contrast to these negative findings, a prospective study of patients undergoing coronary artery bypass surgery (Chandarana et al. 1988) has reported that post-operative cognitive impairment was associated with higher Neuroticism scores on the Eysenck Personality Inventory (EPI), and post-operative perceptual disorders with higher EPI Extraversion scores and 'dominant personality'.

Sensory deprivation

It has been suggested by several authors that impaired vision and hearing predispose to delirium in the elderly (Lipowski 1983). However, little hard evidence has been offered in support of this notion. Rates of visual and auditory impairment increase with age, and the association with delirium may be due to other age-related factors such as dementia or increased vulnerability to physical illness. Several surveys of community elderly populations have reported an association between visual impairment and cognitive impairment (Kay et al. 1964; Maule et al. 1984; Bond 1987; Lindesay 1990). This may be due to impaired performance on information and orientation items of cognitive function scales as a result of inability to read or watch television. Alternatively, as Henderson (1988) has observed, both the brain and the eye are of ectodermal origin, and so both may be vulnerable to similar degenerative changes.

There have been frequent reports of delirium in patients confined in intensive care and coronary care units, and this has been explained in terms of the sensory deprivation experienced in such units. However, Millar (1981) found no relationship between delirium ('post-operative intellectual impairment') and length of time spent in a windowless (theatre recovery or intensive care) room. It seems likely that the barren environment is less important than severe physical illness in the aetiology of delirium in these settings.

Eye operations in the elderly have received particular attention since Ziskind's review (1965) of the relationship between delirium and the practice of covering the eyes with a patch – the so-called 'black patch disease'. He argued that increasing sensory stimulus was of little value, since the problem lay in the patients' perceptual systems. In their review, Summers and Reich (1979) concluded that there was no relationship between age and delirium following such procedures; the use of anticholinergic eye-drops has been proposed as a more important factor (Van Deuren and Missotten 1979).

Adverse life events and difficulties

Delirium associated with traumatic experiences such as bereavement or relocation is well-recognized in the elderly. In some cases, it may be that delirium increases an elderly person's vulnerability to psychosocial stressors (Kennedy 1959), but for the most part this association is probably due to the impact of adverse life events upon individuals at risk of becoming delirious: the demented, the depressed, and the physically ill.

Bereavement

Although there have been no formal studies of the association between bereavement and delirium in the elderly, most reviews of this subject mention bereavement as a potential aetiological factor (e.g. Lipowski 1983), presumably on the basis of personal experience and anecdotal reports. The typical grief reaction following bereavement is characterized by acute pangs of pining, anxiety and restlessness superimposed on a background of more chronic disturbance: dejection, social withdrawal, difficulties with concentration, memory impairment, and disturbances of appetite and sleep (Parkes 1985). This grief and the physical and psychological disturbances associated with it usually subside over time, but in some individuals bereavement is followed by psychiatric disorder. The most common disorders following bereavement are forms of depression or anxiety state, and they are the result of a pathological grief reaction. Other conditions such as alcohol dependence, mania and delirium are less common, and tend to occur in individuals already predisposed to these disorders.

There are several ways in which bereavement might precipitate delirium in vulnerable individuals. The cognitive and vegetative disturbances of typical grief may be a sufficient immediate cause in borderline cases. Alternatively, the death of a partner may remove an important source of care and support; if the surviving spouse's needs are not recognized and met, he or she may become delirious through neglect. In some cases, delirium following bereavement may be iatrogenic, due to the inappropriate 'treatment' of grief with antidepressant and sedative drugs.

Hospital admission and relocation

Goffman's (1968) graphic description of the dehumanizing aspects of admission to institutions is still appropriate to many aspects of admission to hospital (especially of elderly people) today. Although many of the baneful practices he described have been discarded, there is still much in the admission and care of psychogeriatric and elderly medical patients that corresponds with his account of the mortification of the patient's selfhood.

Admission brings the loss of former roles (however tenuous these may appear to an outside observer); the dehumanizing rituals of history taking and examination (especially rectal and vaginal examination); the de-individualizing practice of tagging; the removal of personal possessions; demeaning experiences such as being addressed by uninvited names or wheeled around on a commode in view of others; the restriction of personal space and inability to prevent its invasion; the wearing of clothes stained by previous patients; severe restrictions of autonomy in the choice of company, food, and activity; and, lastly, the exhibition of psychotropic medication to combat what Goffman saw as perfectly understandable defensive reactions to the psychological insult of admission. The principle of 'looping' – when a patient's reaction to mortification leads only to further mortification – is well exemplified by the use of sedative medication, or by the passing of nasogastric feeding tubes in those who appear to choose not to eat.

It would not be surprising if the humiliations visited upon elderly people in some institutions were associated with variations in the manifestations, if not the frequency, of delirious episodes in the elderly. Evans (1987) has described the 'sundown syndrome' in cognitively impaired elderly nursing home residents. This syndrome was identified using a 'Confusion Inventory' which included assessments of behaviour such as wandering, tapping, picking at bedclothes, scratching, rubbing themselves, screaming, and attempting to remove restraints. Subjects were called 'sundowners' if they scored 2.5 or more points higher in the evenings than in the mornings on this Inventory, averaged over two days. Sundowner status was correlated with recency of admission, recency of residence in current room (that is, of relocation), but not with 'morale', nor with several environmental factors such as room illumination or use of restraints. While the 'sundown syndrome' as described in this report probably includes cases of delirium (odour of urine was associated with this label), the description of these subjects suggests that disturbed behaviour in demented residents may well be due to other factors in an institution where physical restraint (including vest and belt restraints) is freely used. Physical restraint may not cause delirium, but studies of intensive care units suggest that the experience of restraint is an important focus for the persecutory hallucinations and delusions reported by delirious patients (Easton and Mackenzie 1988).

The impact of relocation into or between institutions upon the cognitive function of elderly people has yet to be studied in any meaningful detail. As a control group for her study of elderly surgical patients, Blundell (1967) examined the mental state of 16 elderly people before and after admission into an Old People's Home. She found some significant decrements – for example in current affairs memory – but concluded that because of the already deteriorated state of these clients, they were not particularly affected by their move. The numbers were small, and it is not clear how long after first testing they entered the home.

4 The neuronal basis of delirium

Frenesy . . . is an hote postume in certain skynnes and selles of the brayne, and theruppon folowith wakynge and ravynge. And so frensy hath that name frenesis of frenes, felles that beclyppe the brayne. And it cometh in two maners, either of redde colera – chauffed and made lyght with heat of it self . . . or els it cometh of fumosyte and smoke that commethe upwarde'to the brayne, and distourblyth the brayn, and is called Perafrenesi, that is no very frensi.

Bartholomeus Anglicus (1535)

Introduction

Considering the clinical importance of delirium, there has been an extraordinary paucity of research into the neuronal mechanisms underlying it. For the most part our current understanding, suitably translated into the language of metabolism and toxins, has hardly advanced beyond the medieval notions of humours and postumes. So far as delirium in the elderly is concerned, the received wisdom until quite recently was that while some cases were symptomatic, 'senile delirium' was usually the result of degenerative structural changes (Jelliffe and White 1929; Henry 1935; Henderson and Gillespie 1936). Most authors regarded this degenerative senile delirium as a manifestation of arteriosclerotic dementia. Robinson (1939) challenged this prevailing attitude by proposing that all delirium in the elderly was symptomatic, and stated that 'a fixed concept of a sole structural basis of these conditions is not tenable in the light of present knowledge'. Subsequent research and opinion have agreed with him, but while we may know what delirium in the elderly is not, we have still to explain in any detail how clinical delirium at any age is actually caused.

Several authors have attempted to make sense of the various signs and symptoms observed in delirium by explaining them as the diverse manifestations of disorder in a limited number of cerebral processes. The notion of delirium as a disorder of consciousness has been a powerful model for the last hundred years, but as Lipowski (1980b) has argued, this concept is now 'obsolete, vague and redundant', and 'clouding of consciousness' no longer appears as a criterion for delirium in DSM-III-R

(American Psychiatric Association 1987). Lipowski has proposed that the phenomena of delirium are better understood as the results of underlying disorders of cognition and wakefulness. The disorganization of cognitive processes underlies the disorders of thought and perception, and these in turn dictate much of the psychomotor and affective disturbance. Similarly, disordered wakefulness with attentional deficits and disturbance of the sleep–wake cycle result in distractability, field restriction, impaired registration of new memories, daytime drowsiness and nocturnal insomnia. It has been suggested that attentional deficits are the cause of cognitive disorders in delirium (Geschwind 1982; Mesulam 1985), and that dreams and oneiric thinking are the source of delirious hallucinations and delusions; Hunter (1835) described delirium as 'a diseased dream arising from what may be called diseased sleep'. As Lipowski (1980b) has pointed out, the advantage of conceptualizing delirium in terms of underlying disorders of cognition and wakefulness is that it allows particular aetiological hypotheses to be constructed and tested against the evidence.

This chapter will review the current evidence and opinion concerning the neuronal basis of delirium. Much of what follows is speculative and hypothetical, and is presented here not as a summary of what is known but as a stimulus for much-needed research. The neuronal mechanisms of delirium deserve serious study, since they are likely to illuminate other areas of current interest in neuropharmacology, neurophysiology, and neuropsychology.

Delirium associated with metabolic disturbance

Clinically, the delirium syndrome represents a relatively limited response to a wide range of insults, and various causal theories have been propounded to explain this. Underlying most of them is the concept of altered cerebral metabolism, in particular the reduction of oxidative metabolic processes. The causes of this altered cerebral metabolism include hypoxic or hypoglycaemic substrate deficiencies (Siesjo et al. 1976), vitamin deficiencies, the poisoning of the cerebral metabolic pathways by toxins (perhaps as a result of impaired hepatic detoxification processes), and the possible accumulation of toxic 'false transmitters' by neuronal uptake mechanisms (James et al. 1979). Disturbances in cerebral blood flow and increased permeability of the blood–brain barrier to toxins and drugs may also contribute to disordered cerebral metabolism in some cases.

Impaired cerebral metabolism probably causes clinical delirium by disrupting synaptic transmission between neurons and increasing the levels of 'neural noise' (Grimley Evans 1982). If some components of the CNS are more vulnerable to metabolic disturbance than others, it may be that delirium is not necessarily the result of a 'widespread derangement of cerebral metabolism' (Lipowski 1980b), but is due to the disruption of specific neuronal systems. This possibility is supported by the EEG evidence, since the bisynchronous delta waves observed in metabolic delirium are also seen following circumscribed structural lesions in the brainstem and diencephalon, and are thought to reflect disruption of deep central structures, probably in the reticular formation (Gloor *et al.* 1977; Muller and Schwartz 1978; Spehlmann 1981).

The cholinergic hypothesis

Blass and Plum (1983) have suggested that the reduced synthesis of certain neurotransmitters, notably acetylcholine, is the final common pathway for delirium. If so, this is not due to a general impairment of energy production in the brain, since neurotransmitter synthesis consumes negligible quantities of ATP (Bachelard 1975). A more likely cause of the neurotransmitter failure associated with disturbed brain metabolism is substrate deficiency. Acetylcholine synthesis may be particularly vulnerable in this respect, since one of its immediate precursors, acetyl-coenzyme A, is also a vital and potentially rate-limiting component of the aerobic citric acid cycle (Fig. 4.1). Insulin-induced hypoglycaemia has been shown to reduce the acetylcholine content of the brain in animals (Crossland *et al.* 1955). Hepatic encephalopathy may be partially caused by reduced rates of acetylcholine synthesis following diversion of the alpha-ketoglutarate component of the citric acid cycle to glutamate and glutamine by toxic levels of ammonia (Fig. 4.1) (Sherlock 1985). Vitamin deficiencies also reduce the synthesis of neurotransmitters; for example thiamine deficiency in animals results in reduced cerebral concentrations of acetyl-coenzyme A and acetylcholine (Heinrich *et al.* 1973).

The notion that the disturbance of central cholinergic mechanisms is an important cause of delirium is supported by the clinical observation that anticholinergic drugs such as atropine, procyclidine, and amitryptiline that can cross the blood–brain barrier are potent precipitators of delirium. The elderly are particularly vulnerable to this effect of anticholinergic drugs (Blazer *et al.* 1983), as are patients with dementia of the Alzheimer's type (DAT) (Sunderland *et al.* 1987). The delirium associated with these drugs can be temporarily reversed by the administration of the anticholinesterase inhibitor physostigmine (Duvoisin and Katz 1968; Greene 1971; Granacher and Baldessarini 1975; Milam and Bennett 1987). There is a large body of evidence concerning the psychotropic effects of anticholinergic drugs, not least because they have been regarded by behavioural pharmacologists as a 'model psychosis' for schizophrenia (Warburton and Wesnes 1985a), and as a model for DAT. However, it would seem more appropriate to regard

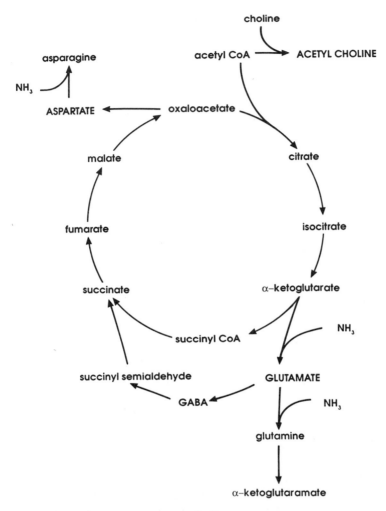

Fig. 4.1 The citric acid cycle.

these findings as evidence about the mechanisms underlying delirium, since delirium is what anticholinergic drugs actually cause in vulnerable cases. Studies of healthy volunteers receiving large doses of anticholinergic drugs also show significant impairments in orientation, selective attention, information processing and abstract thinking (Singh and Kay 1985). Anticholinergic drugs also impair memory function, in particular the encoding of short-term memories into the long-term store (Richardson *et al.* 1985).

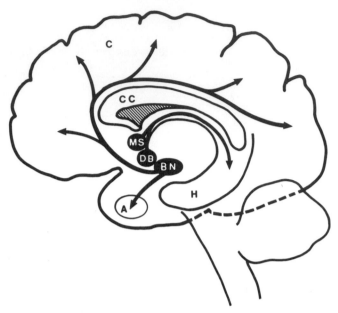

Fig. 4.2   Principal cholinergic projections in the brain: A, amygdala; BN, basal nucleus of Meynert; C, cortex; CC, corpus callosum; DB, nucleus of the diagonal band; H, hippocampus; MS, medial septum.

Attention and arousal

If delirium is caused by impaired central cholinergic function, what is the evidence that cholinergic systems are involved in processes such as attention and the sleep–wake cycle that are characteristically disrupted in this condition? Several central cholinergic pathways have been identified (Robinson 1985), and there is a large body of animal behaviour research and neurochemical evidence that these are involved in the regulation of attention and sleep. The principal cholinergic projection to the neocortex arises from neurons in the basal nucleus of Meynert (Fig. 4.2). Another well-described central cholinergic system projects from cell bodies in the medial septum and nucleus of the diagonal band to the hippocampus, the so-called septo-hippocampal pathway (Fig. 4.2). In their review of evidence concerning the relationship between central cholinergic systems and behaviour, Warburton and Wesnes (1985*b*) conclude that a cholinergic system is involved in an input-processing neuronal mechanism that focuses attention by selecting from the various internal and external stimuli those that will determine relevant behavioural responses. They argue that an ascending excitatory cholinergic reticulocortical pathway amplifies evoked potentials at the cortex, and that this results in a more efficient selection of the subset of sensory inputs that determine subsequent responses. On the output side of the behavioural chain, the cholinergic component of the septo-hippocampal system may also be involved in facilitating selective attention by inhibiting inappropriate and unrewarded responses (Singh 1985).

The sleep–wake cycle

Disrupted cholinergic transmission may also underlie the characteristic disturbances of the sleep–wake cycle in delirium, since cholinergic systems appear to form part of the neuronal mechanism responsible for this cycle. At the level of the cortex there is increased acetylcholine release during wakefulness and REM sleep (Jasper and Tessier 1971). However, the role of cholinergic neurons in the sleep–wake cycle does not appear to be restricted to their neocortical activity, since excitation experiments indicate that acetylcholine can also induce slow-wave sleep, particularly if applied to certain circuits involving the limbic cortex, forebrain, and midbrain (Hernandez-Peon *et al.* 1963). Furthermore, cholinergic mechanisms also appear to be involved in the organization of REM sleep.

The conclusion that may be drawn from numerous lesion and stimulation studies in animals is that the waking state is maintained by cells in the midbrain portion of the reticular formation (MRF) (Steriade 1983). The MRF neurons involved do not project directly to the cortex, but synapse in the medial intralaminar thalamic nuclei. Lesions of these nuclei result in a pronounced alteration of consciousness (see page 75).

Attempts to identify discrete centres involved in the induction of slow-wave and REM sleep have been less successful, and it is likely that the various phases of the sleep–wake cycle are the result of co-ordinated activity in a number of distinct neuronal and neurotransmitter systems, of which the tonic 'waking centre' in the MRF is but one (Lindsley 1983). There is evidence that projection neurons in the basal forebrain are involved in the production of slow-wave sleep (McGinty and Szymusiak 1989). The basal forebrain is an anatomically complex region, and areas implicated in hypnogenesis have included the medial and lateral preoptic areas, and the substantia innominata which contains the basal nucleus of Meynert. As noted above, this nucleus is of interest in the context of a cholinergic hypothesis of delirium because it is the origin of the major cholinergic projection to the cortex. However, it is unlikely that the reduction in cholinergic activity observed in the cortex during slow-wave sleep is due to inhibition of the direct cortical projection of this nucleus, since the severe depletion of this pathway that occurs in DAT is not associated with hypersomnia.

Basal forebrain structures also project to the MRF and may influence arousal by this means. The thalamus may also be a site of regulation of the sleep–wake cycle; there is a cholinergic pathway from the basal nucleus to the thalamic reticular nucleus (Levey *et al.* 1987), and there are reciprocal projections from the magnocellular portion of the dorsomedial thalamic nucleus to the substantia innominata (Von Cramon *et al.* 1985). The notion that the thalamus has an important role in the regulation of sleep was emphasized by Hess (1969), but this idea was later neglected in favour of the hypothalamus and reticular activating systems as sleep-regulating centres.

REM sleep is not a single entity, but is composed of various dissociable elements mediated by various neuronal groups within the pontine tegmentum. These elements include the rapid eye movements themselves, the characteristic ponto-geniculo-occipital (PGO) waves, and tonic atonia of the anti-gravity muscles. Monoaminergic neurons in the locus coeruleus and the dorsal raphe show reduced activity in REM sleep, and may be involved in the 'switch' from slow-wave to REM sleep and in the release of PGO waves. There is also evidence that a population of possibly cholinergic neurons in the region of the locus coeruleus specifically increase their activity during REM sleep, and mediate the associated muscular atonia via a ponto-medullary-reticulospinal pathway (Sakai 1980). Disruption of this pathway produces orienting and rage behaviour during REM sleep in animals, and it may be that a similar dysfunction in humans contributes to nocturnal disturbance in delirium.

Certain aspects of the sleep–wake cycle are known to alter with age (Morgan 1987), and it may be that such changes in the neuronal mechanisms controlling the sleep–wake cycle increase their vulnerability to disturbance. In particular, there is an age-related increase in the number of shifts between sleep stages, and in the number of nocturnal awakenings, which suggests that the control mechanism becomes less stable over time.

Other central cholinergic systems

Cholinergic mechanisms centred on the basolateral amygdala have been implicated in the expression and control of aggression (Bell and Brown 1985). Stimulation of these areas in animals elicits a defensive response with increased arousal and orienting behaviour, but it also inhibits attack. As Singh (1985) has observed, the function of the cholinergic system within the amygdala circuit appears to be comparable to that of cortical cholinergic projections in that it increases attention while inhibiting automatic aggressive responses. Disruption of such a system in humans might account for the disinhibited aggressive behaviour that occurs during delirium.

Similarly, the disorders of visual perception associated with delirium may be caused by disturbed central cholinergic function, since the visual system appears to have significant cholinergic components at all levels. In addition to the cholinergic input to the visual cortex, there is evidence from animal studies of cholinergic innervation in the superior colliculus, lateral geniculate body (Domino *et al.* 1973), and retina (Neal 1976).

Monoaminergic dysfunction and delirium

The cortical arousal system appears to be composed of several functional components, of which the cholinergic component subserving attentional control is only a part. Monoaminergic projections from the reticular formation are also important in the maintenance of arousal, and the two systems may interact, with the cholinergic mechanism increasing attention while selectively inhibiting the more automatic motivational and affective responses to stimuli mediated by the monoaminergic arousal system (Singh 1985). Central monoaminergic systems also appear to be involved in the control of the sleep–wake cycle, as indicated above.

The delirium secondary to alcohol withdrawal is associated with increased cerebrospinal fluid (CSF) levels of noradrenaline (NA) (Hawley *et al.* 1981) and its metabolite 3-methoxy,4-hydroxyphenylglycol (MHPG) (Fujimoto *et al.* 1983). Raised levels of 5-hydroxyindoleacetic acid (5HIAA) in the CSF and reduced levels of 5-hydroxytryptamine (5HT) in the plasma have also been reported in association with delirium tremens and clozapine withdrawal (Banks and Vojnik 1978). Modai *et al.* (1986) have reported a case of delirium following sudden phenelzine withdrawal associated with increased levels of urinary MHPG.

Delirious states associated with withdrawal may be due to chronic changes in neuronal receptor populations rather than to acute metabolic disturbances; alcohol withdrawal has been shown to be associated with increased central beta-adrenergic receptor sensitivity (Hawley *et al.* 1981). Drugs acting on monoaminergic receptors have been associated with delirium; however, their effects are probably rather complex since they also inhibit acetylcholine release from presynaptic neurons (Dilsaver *et al.* 1987). Propanolol has been reported both as a cause (Kuhr 1979) and a cure (Petrie and Ban 1981) of delirium.

Dysfunction of monoaminergic neurotransmitter systems may also contribute to hepatic encephalopathy. Circulating levels of the aromatic amino acids tyrosine and tryptophan (the precursors of monoamine neurotransmitters) are increased in liver disease, perhaps as a result of impaired hepatic deamination (Sherlock 1985). This probably does not alter the rate of synthesis of noradrenaline and dopamine, since the proportion of tyrosine used by the brain for this purpose is minimal. However, the rate of synthesis of 5HT is affected by changes in the level of plasma tryptophan. Another result of defective hepatic deamination is the increased production of false neurotransmitters. The enzyme dopamine-beta-hydroxylase is relatively substrate-non-specific, and will convert most phenylethylamines to their corresponding phenylethanolamine; for example, it will convert tyramine to octopamine, and phenylalanine to beta-phenylethanolamine. In patients with a significant porto-caval shunt, substances such as tyramine and octopamine produced by bacteria in the gut are another source of false neurotransmitters. Fischer and Baldessarini (1971) have reported an improvement in the mental state of patients with hepatic encephalopathy following treatment with levodopa, suggesting that it may act by reversing the accumulation of false neurotransmitters. However, it has yet to be demonstrated that the the uptake of false neurotransmitters by central neurons is specifically associated with delirium in liver failure.

Other central neurotransmitters and delirium

Only a small proportion of the neurons in the brain are cholinergic or monoaminergic, and it is possible that impairment in the function of other neurotransmitter systems also contribute to the clinical picture of delirium. In particular, the neurotransmitters glutamate and gamma-aminobutyric acid (GABA) may be involved, since glutamate is a major excitatory neurotransmitter in the brain, and GABA a major inhibitor. Both substances are intimately involved in the intermediary metabolism of the cell (Fig. 4.1), and are therefore vulnerable to a wide range of metabolic disturbances.

The roles of glutamate and GABA in delirium have hardly been studied, but there is some evidence from investigations of liver failure and hepatic encephalopathy which suggests that they play a part (Schafer and Jones 1982). GABA is produced by the gut flora, but this is normally prevented from reaching the brain by liver metabolism and the blood–brain barrier. In liver failure the combined effect of porto-caval shunting and increased blood–brain-barrier permeability may result in increased central GABA-ergic neuronal inhibition. In animal models of hepatic coma, there is evidence of increased post-synaptic GABA binding (Baraldi and Zeneroli 1982), and decreased binding of glutamate. Sedative withdrawal is associated with a transient decrease of GABA function (Haefely 1986), which may account for the associated increase in central excitability and convulsive phenomena such as myoclonus and fits observed in withdrawal delirium.

There may also be dysfunction of peptide neurotransmitters in delirium. Koponen et al. (1989c) have found a statistically significant decrease in beta-endorphin-like immunoreactivity in the CSF of elderly delirious patients compared with non-delirious controls. The lowest levels were found in patients with hypoactive delirium. The reason for this association is not known, but it may be that beta-endorphin dysfunction increases susceptibility to delirium. Beta-endorphins appear to be involved in the inhibitory control of cortical cholinergic function (Moroni *et al.* 1978), and of the locus coeruleus. Koponen *et al.* (1989*d*) have also demonstrated reduced somatostatin-like immunoreactivity associated with delirium in their patients. It should be borne in mind that a large proportion of their delirious sample was also demented, so it may be that the abnormal CSF findings are related to the dementia rather than to the delirium.

Another substance that may be involved in the pathogenesis of delirium is the endogenous pyrogen interleukin-1, which is released from cells following a wide range of infectious, inflammatory and toxic insults. Interleukin-1 is a promoter of slow-wave sleep in animals (Krueger *et al.* 1984), and a similar action in humans may contribute to the sleep disturbance observed in delirium.

Delirium associated with structural lesions

In humans, evidence for the location of specific areas of the brain involved in the aetiology of delirium is derived from the case reports and clinical series of patients affected by particular lesions, and from the rapidly developing techniques of neuroradiological brain imaging. A clinical picture very similar to that of delirium secondary to other causes has been reported following certain thalamic lesions. In particular, the so-called 'medial thalamic syndrome' is characterized by attentional deficits, sleep disturbance, mood change, autonomic disturbance, and cognitive impairment (Martin 1969). This syndrome usually comes about as the result of infarction or haemorrhage. Diffuse encephalopathy has been described following non-dominant thalamic infarction (Friedman 1985), and Santamaria *et al.* (1984) describe a 'confusional syndrome' associated with haemorrhagic anteromedial thalamic strokes. Their patients developed attentional deficits, disorientation in time and place, anterograde memory loss, agitation, disinhibition, urinary incontinence, and asterixis. They comment that such strokes should be considered as a differential diagnosis in cases of delirium.

It is difficult to associate clinical syndromes with specific haemorrhagic loci in thalamic strokes, since these are rarely confined to specific nuclei and also involve intrathalamic tracts such as the mamillothalamic tract and the lamina medulla interna. However, Lugaresi *et al.* (1986) have reported a detailed case history of a specific familial degeneration of the thalamus confined to the magnocellular portion of the dorsomedial nuclei that produced many symptoms reminiscent of delirium, including insomnia, intrusive dream imagery, a fluctuating attention deficit proceeding to stupor, temporospatial disorientation, autonomic disturbance, and myoclonus. There was a progressive slowing and flattening of the EEG with the absence of characteristic sleep features. It was also noted that atropine caused a substantial deterioration in the patient's condition, suggesting that disruption of a cholinergic system may have been involved in the aetiology of the clinical syndrome.

The thalamus is not the only area of the brain that, when damaged, results in a delirium-like state. Right (non-dominant) hemisphere infarcts in the territory of the middle cerebral artery also characteristically present as 'acute confusional states' with attentional deficits and reduced coherence of thought and action (Mesulam *et al.* 1976). It has been suggested that this association between right hemisphere infarcts and delirium reflects a right hemisphere specialization in the regulation of attentional tone, at least in dextrals. This association of attentional mechanisms with the non-dominant hemisphere raises an interesting possibility with regard to our current practice of cognitive testing in delirium. Standardized brief cognitive assessment instruments such as the MMSE are relatively insensitive to right hemisphere deficits, since they have very few items dealing with non-dominant visuospatial functions (Dick *et al.* 1984; Mori and Yamadori 1987). If more visuospatial items were to be included, would these instruments be more sensitive in identifying delirium?

In a review of the various unilateral and generalized neglect syndromes associated with CNS lesions in monkeys and humans, Mesulam (1981) has proposed a model of directed attention co-ordinated by an integrated network of four cerebral regions. He defines attention as a complex higher-order function with seperately localized component processes as follows:

(1) Posterior parietal neocortex: this area contains an internal sensory map, together with a mechanism for modifying the extent of synaptic space devoted to specific portions of the external world. It may be that disturbance of this latter function causes the progressive restriction of the attentional field observed in delirium.

(2) Limbic structures: according to Mesulam these determine the 'spatial distribution of motivational valence'.

(3) Frontal cortex: neuronal systems located here subserve the exploratory/scanning aspects of attention.

(4) Reticular systems: these are responsible for the tonic maintainence of arousal and vigilence. In the monkey, the relevant nuclei identified by

Mesulam are the intralaminar thalamic nuclei, the locus coeruleus and the midline raphe nuclei.

There is also some evidence that different anatomical lesions are associated with different clinical types of delirium. Mori and Yamadori (1987) have studied a sample of 41 subjects with infarcts in the territory of the right middle cerebral artery. Of these, 25 developed an 'acute confusional state' (ACS), and six an 'acute agitated delirium' (AAD) according to their diagnostic criteria. ACS was associated with lesions involving the inferior and mid-frontal gyrus and basal ganglia on CT scanning, whereas AAD was associated with lesions involving the mid-temporal gyrus. There was no evidence of significant involvement of the parietal areas in this study; the authors suggest that right parietal lesions may cause unilateral neglect (a directed attention deficit), but right frontostriatal lesions cause bilateral neglect (a global attenion deficit) that manifests as an 'acute confusional state'. They also suggest that ACS is due to attentional deficits, in contrast to AAD, which is caused by the disruption of a cortical association area for sensory-limbic interaction (the upper mid-temporal gyrus). In support of this, Caplan et al. (1986) have reported AAD developing in 5/10 subjects with lesions in the inferior right middle cerebral artery territory involving the temporal lobes. Similarly, agitated delirium associated with homonymous hemianopia and visual hallucinations has been described following lesions of the medial temporo-occipital region (Medina et al. 1977).

Evidence from brain imaging of unsedated delirious patients is limited by several factors. In the first place, only those who are in a position to co-operate with the procedures involved will be studied; second, most studies to date have been confined to neurological samples; and third, lesion location on CT scanning does not always coincide with the anatomical extent of the lesion. Swigar et al. (1985) correlated CT scan findings with psychopathological features in a sample of 50 elderly patients with a range of organic and functional disorders. This was not a study of delirium, but clinical features indicative of delirium such as 'confusion/perplexity', 'disorientation', and a fluctuating mental state were associated with left-sided parieto-temporal atrophy. Koponen et al. (1989a) compared 69 elderly neurological patients with DSM-III delirium with a group of non-delirious neurological controls. They also found cortical atrophy and increased ventricular dilatation in the delirious patients. They found that delirious patients also had an increased rate of focal lesions, and in contrast to Swigar et al. (1985) these were predominantly in the right hemisphere cortical association areas. Hyperactive delirium was associated with right parieto-occipital lesions.

Functional imaging techniques such as positron emission tomography or

nuclear magnetic resonance imaging have not yet provided much information about the neuronal basis of delirium. In an investigation of the
regional cerebral blood flow of 12 non-elderly patients with delirium
tremens due to alcohol withdrawal using xenon-133 inhalation tomography, Hemmingsen *et al.* (1988) found that visual hallucinations and
agitation were significantly associated with increased cerebral blood flow.
Functional imaging has yet to be applied to the hypoactive delirium that is
more typical in the elderly.

Delirium associated with psychological trauma

The mechanisms by which traumatic experiences such as bereavement and
relocation may precipitate delirium in certain elderly people in the absence
of any specific physical cause remain entirely speculative. It may be that
individuals who respond in this way to such stresses are particularly
vulnerable to delirium, and relatively minor metabolic disturbances are
sufficient to precipitate the syndrome. Alternatively, the stress response
itself may be a sufficient cause of delirium in some cases. Kral (1975) has
proposed that delirium in the elderly might be an acute stress reaction
mediated by corticosteroids, suggesting that this response is due to 'age-
linked weakness of the stress-resisting mechanism' in the brainstem and
the hypothalamus. Circulating catecholamines are also elevated in stressed
individuals, and Lipowski (1983) has suggested that the resulting accentua-
tion of cerebral metabolic demands may result in delirium in the elderly,
particularly the demented. It has also been suggested that these peripheral
catecholamines may precipitate delirium in the elderly by crossing the
blood–brain barrier, or altering cerebral blood flow to particular cortical
regions (see page 41). Reduced levels of plasma cortisol have also been
reported as a cause of cognitive impairment and delirium in old age
(Basavaraju and Phillips 1989).

Conclusions

If delirium is principally caused by the failure of specific neuronal systems
involved in the regulation of attention, the sleep–wake cycle, aggression,
and the visual system, do different chemical and structural lesions cause
different types of delirium? As we have seen, neurologists distinguish
between states of 'acute confusion' in which the deficits are primarily
cognitive, and 'delirium' in which cognitive impairment is accompanied by
delusions, hallucinations, autonomic arousal, altered affect, and behavioural disturbance. Psychiatrists on the other hand tend to ignore this

distinction, and it is is certainly true that cases of delirium can be found to occupy all points on the continuum between the extremes of hypoactivity and hyperactivity. Nevertheless, the EEG evidence suggests that there are at least two types of delirium: the typically hypoactive delirium associated with slowing of the EEG, and the hyperactive delirium associated with withdrawal states in which the EEG shows no abnormality apart from paroxysmal bursts of fast activity (Pro and Wells 1977). There is also evidence that patients with hypoactive delirium are more severely cognitively impaired than those with hyperactive or mixed states (Koponen *et al.* 1989*b*). Itil and Fink (1966) have proposed that clinical delirium is due to an imbalance between central cholinergic and adrenergic mechanisms. It may be that pure 'cholinergic' delirium and pure 'monoaminergic' delirium represent two extreme types of the disorder, and one hypothesis that deserves to be investigated is that hypoactive delirium is caused by cholinergic dysfunction, hyperactive delirium by monoaminergic dysfunction and mixed states to a combination of the two. The contribution of disordered transmission in other central neuronal pathways also needs to be studied.

5 The management of delirium

Note that you must eschew continuall use of stupefactive medicines as well inwardly as outwardly also. For in this disease, by overmuch cooling you may turne the frenesy into a litargy, wherby you may cause him to sleepe so, that you can awake him noe more. Also if the patient be weake beware how you minister stupefactive things to provoke slepe, for in such as be weak (as Trallianus saith) somnoriferous potions do noe small hurt, and sometime they kill.

Barrough (1583)

There are two essential aims in the management of delirium in the elderly: treatment of the underlying causes, and control of the patient's distress and disturbed behaviour. However, as Grimley Evans (1982) has observed, these aims are often in conflict in hospital settings, and careful thought needs to be given at the outset to where the patient will be most effectively managed.

At home or in hospital?

The decision to admit an elderly person with delirium to hospital should not be taken lightly, since familiar routines and established networks of care will be disrupted, and the delirious symptoms and distress will almost certainly worsen as a result. Following a domiciliary assessment of the patient's history and physical state, the history from relatives, and any available district nursing records, it is often possible to make a realistic first guess as to the cause of the delirium. Treatment should be started at once, and blood, urine, and possibly sputum samples should be taken. The blood sample should be checked for evidence of electrolyte imbalance, renal or hepatic impairment, a possible silent myocardial infarction, previously unsuspected diabetes, or toxic levels of any relevant drug, such as phenytoin. It is prudent to involve the patient's general practitioner and the district nursing service if treatment is started at home, since they can be very useful in the monitoring of progress.

Causes of delirium such as drug toxicity and infection can often be sucessfully treated at home. The management of myocardial infarction at home is still the subject of some contention, but it is probable that if adequate support is available and there is no apparent shock, arrhythmia, or heart failure, then the prognosis of elderly patients is no worse than if they are moved to hospital (Hill *et al.* 1979).

The benefits of a familiar home environment must be balanced against the greater difficulties of physical examination, investigation and treatment of the underlying cause. Generally speaking, if the patient is seriously ill, if the cause of the delirium cannot be found, if there is no satisfactory improvement after 24 hours, if there is no one available to care for the patient, or if disturbed behaviour makes care at home impossible for the family, then admission to hospital is indicated. Likewise, a delirious patient who presents to the casualty department should also be admitted, since the additional trauma of the return home usually makes matters worse. The decision whether to admit or not should always involve the patient's general practitioner, who is in the best position to weigh the particular risks and benefits.

For the elderly delirious patient in hospital, the balance of risks and benefits is reversed. It would seem at first sight that the management of delirious patients in hospital must be better than management at home by any criteria. It is certainly true that biochemical abnormalities can be rapidly reported to the physician in charge, and a greater range of investigations is available. Once started, treatment can be monitored more closely to ensure that both the underlying disorder and the delirium are resolving without undue delay. A number of different but possibly equally contributing causes can be treated rapidly and simultaneously, and any complications of the treatment can be detected promptly. However, some of these advantages may be more apparent than real, and a note of caution must be sounded. First, the fact that a patient is in hospital tends to lead to over- rather than under-investigation, particularly if the history and examination are hampered by the delirium. A number of tests may be ordered that would be unavailable at home (chest X-ray, ECG, possibly abdominal ultrasound, skull X-ray, lumbar puncture), and these add to the physical and mental trauma, especially if moves around the hospital are required. If the causes of delirium can be diagnosed and treated clinically, the indications for some of these tests may be better reviewed at a later date, and carried out when the patient is more stable.

Second, all of the abnormalities revealed by the investigations tend to be treated vigorously in hospital. Although this may be appropriate in some cases, it is not always true that the first abnormal laboratory result to be returned is indicative of the cause of the delirium. The importance of a clinical diagnosis of the underlying causes cannot be over-emphasized. The finding of a haemoglobin level of 9.0 g/dl in an aggressively delirious patient does not necessarily require immediate transfusion, and it is more likely that the delirium is primarily due to another cause. The trauma of

an intravenous line, together with intrusive half-hourly nursing observations would be very disturbing to such a patient, who would then require
sedation, perhaps avoidable otherwise.

Third, vigorous treatment, even if successful, does not necessarily hasten
the departure of the delirium in any given patient, and over-rapid
treatment may itself be harmful. For example, very rapid intravenous
rehydration of a hyperosmolar diabetic patient *may* be necessary if there
is evidence of cardiovascular collapse or decreasing renal function, but it
may worsen the delirium by precipitating cerebral oedema. A further
round of investigations for unresolved delirium may then be undertaken,
delaying still further the reorientation of the patient to the new
environment.

Finally, patients in hospital have to conform to the needs of the new
ward community. This will probably involve a significant disruption of
sleep–wake pattern, the appearance of a large number of strangers at
unpredictable times, and an apparent requirement for absolute silence at
night and continence at all times. As a result, sedation will almost
certainly be given – usually on an 'as required' (in other words a 'too
late') basis, there is risk of urinary catheterization, and hydration will
almost automatically be through a vein rather than by a familiar person
bringing cups of tea. Much of this may be unavoidable, but if the cause
of the delirium can be found and treated it may well be that the most
stormy period can be weathered without additional insults, provided
there is tolerance, time available, and an understanding of the nature of
the disorder.

It is important that the potential advantages of in-patient investigation
and treatment are not negated by inexpert management. An inexperienced
junior doctor attempting to manage a life-threatening cardiac arrhythmia
without advice would (or should) be corrected in his ways, and expert help
should likewise be sought in the management of delirium. All too often,
the vicious circle of delirium, disturbed behaviour, and sedation experienced by elderly patients in hospital is the result of poor knowledge and
attitudes of the medical and nursing staff; it could be avoided altogether
by adequate training in the early detection and non-pharmacological
control of delirium.

Case 4

A 78-year old widow with a moderate cerebrovascular dementia developed a post-
operative delirium, apparently due to digitalis toxicity and hypokalaemia. Despite
treatment, her delirium failed to respond, and extensive medical examination and
investigation revealed no other obvious organic cause for her continuing delirium.
She was removed from the surgical ward by her relatives, to the nursing staff's
disquiet. Her conscious level improved markedly in the car park, and after the ten

minute drive home it had improved further. Her consciousness was clear within half an hour of arrival at her home. Two days later she took a half-mile walk, and was able to be left at home unaccompanied the following night, returning entirely to her pre-operative level of functioning.

Sometimes the underlying cause of the delirium cannot be identified because the patient is either too disturbed or too sedated. Such cases may be managed best on a specialist acute psychogeriatric assessment ward, provided that medical advice is readily avaiable. Disturbed behaviour is better tolerated and understood in this setting, and this allows the vicious circle of delirium and sedation to be interrupted. In some cases, however, psychiatrists and psychiatric nurses will not be able to manage the physical disorders adequately, and in a perfect world they would be cared for on a joint geriatric/psychogeriatric acute assessment unit. Unfortunately, few such units exist as yet.

Involuntary admission to hospital

Occasionally, elderly people with delirium refuse to leave home in order to go into hospital. If the patient can be managed at home, or is clearly in the terminal phase of an incurable illness, then this wish should be respected. Problems arise if the cause of the delirium is unknown and requires hospital investigation, or if treatment in hospital is necessary to save life. In the UK, the doctor's common law duty of care in such cases may require quick removal to hospital without any formal procedures. Where possible, however, delirious patients should be formally admitted to hospital under the 1983 Mental Health Act. The powers granted by this Act are often not used in these cases because clinicians are unsure whether or not delirium is a 'mental disorder', as required by the Act. In fact, the term 'mental illness' (the relevant category of mental disorder) is not defined anywhere in the Act, and the decision is entirely one of clinical judgement. Admission to hospital under Section 2 is usually the most appropriate course, as this allows for both assessment and treatment of the delirium. Patients do not need to be a danger to others before they can be detained; it is sufficient that they require admission in the interests of their own health and safety. The existing legislation to detain delirious individuals may not be used because physicians are reluctant to admit formal patients to general medical beds, and psychiatrists are equally unwilling to admit seriously ill elderly people to mental hospitals where the necessary medical and surgical resources may be lacking.

Another piece of legislation that is sometimes used to remove elderly people to hospital in the UK is Section 47 of the 1948 National Assistance Act. To meet the criteria of this Section, patients must be:

(1)   (a) suffering from grave chronic disease; or
      (b) being aged, infirm or physically incapacitated, are living in
           insanitary condition;
and
(2)   (a) are unable to devote to themselves, and
      (b) are not receiving from others, proper care and attention.

Section 47 provides none of the reviews and safeguards of the 1983
Mental Health Act, nor does it allow treatment against the patient's
wishes, and it should not be used to remove elderly people with mental
disorders, including delirium (Wolfson *et al.* 1990).

Case 5

An 83-year old widow, living alone and with no close relatives, with an unclear
history of cognitive decline in the setting of chronic congestive cardiac failure, was
referred by her general practitioner as an emergency to the psychogeriatric service
with a two week history of nocturnal shouting, sudden decline in self-care,
worsening of incontinence, and visual hallucinations. She was immobilized by
severe cellulitis of both oedematous legs. She refused adamantly to go into
hospital, resisting violently any such suggestion. Options considered included an
admission under Section 47 of the National Assistance Act, and compulsory
admission, without formality, on common-law grounds. The final decision, taken
jointly by the general practitioner, consultant psychogeriatrician and social worker,
was to regard the delirium as a 'mental disorder' in legal terms, and to admit her
compulsorily under the 1983 Mental Health Act to a medical ward in a hospital
which also included a psychiatric unit. This was the first such admission to a non-
psychiatric ward of this hospital. The patient responded well to antibiotic treatment, and a severe dementia was revealed when clear consciousness returned. She
was eventually transferred to a residential home. There was considerable confusion
amongst nurses and doctors on the medical ward as to the patient's rights, and
their responsibilities; none had any familiarity with the Mental Health Act
whatsoever. The Mental Health Act Commissioners were apparently equally
nonplussed by the lack of precedent.

A related management problem is presented by the delirious in-patient
who suddenly and irrationally expresses a wish to leave hospital. All too
often, this problem arises in the middle of the night with the patient
making a bid for freedom, and it has to be managed by relatively junior
and inexperienced medical and nursing staff, some (if not all) of whom are
unfamiliar with the case. The principles of management in these circumstances are straightforward. In an emergency, the doctors' and nurses'
common law duty of care will justify any reasonable restraint and sedation
necessary to detain a physically ill and delirious patient (DHSS 1976).
Once the situation has been brought under control, however, it is good

practice to arrange for the patient to be detained formally under the appropriate Section of the 1983 Mental Health Act if this behaviour is likely to be repeated.

The Mental Health Act also provides for patients to be detained for up to six hours by a psychiatrically trained nurse, pending the arrival of a doctor. Section 5 (4) allows the detention of 'a patient already receiving treatment for a mental disorder in hospital', provided that:

(1) the patient is suffering from mental disorder to such a degree that it is necessary for his health or safety, or for the protection of others for him to be immediately restrained from leaving the hospital;

and

(2) that it is not practicable to secure the immediate attendance of practitioner for the purpose of formulating a report under Section 5(2) (emergency detention by a medical practitioner).

Although this Section can be used in any ward of any hospital, its utility on non-psychiatric units is often limited by the stipulations that the patient must already be receiving treatment for a mental disorder, and that the detaining nurse should have the appropriate psychiatric training. Any medical, surgical, or geriatric units wishing to make use of this particular provision of the Mental Health Act should first institute a regular delirium screening policy; provided that the patient has been identified as delirious prior to the emergency, any subsequent treatment of the cause of the delirium should satisfy both the letter and the spirit of the Act. Second, they should either employ nurses with the necessary psychiatric experience, or else they should ensure close liaison between their nursing staff and those of any psychiatric units on the same hospital site.

Physical apsects of management

What and when to investigate

This section sets out some basic principles of investigation in delirium. However, it should be remembered at all times that any attempt to lay down strict rules is to deny another basic principle, namely that each patient is different and should be treated accordingly. Complex algorithms (Fig. 5.1) are useful, but they are no substitute for clinical acumen.

Clinical examination of the patient in hospital includes regular observations by nurses of temperature; pulse rate (and apex rate if irregular);

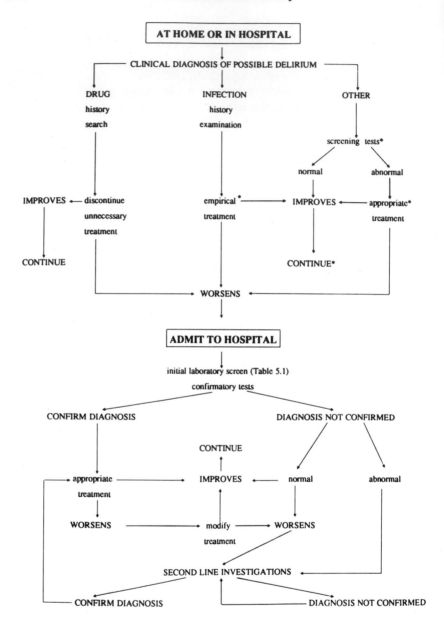

* *May be carried out at home only if support is good.*

Fig. 5.1 The investigation and management of delirium.

blood pressure; respiration rate; the occurrence of vomiting and diarrhoea; oliguria; constipation; fits; blood sugar monitoring, if the patient is known to be diabetic; sputum production; the presence of offensive urine; urine tests for protein, blood, and glucose; and perhaps fluctuation and deterioration of conscious level. These observations may be vital to establish the cause of the delirium, and may be missed by the most thorough doctor on a single examination. A specific request should therefore be made for such observations to be carried out.

The clinical assessment by the doctor must include careful documentation of cardiovascular status including the presence of cardiac failure and murmurs; any abnormal neurological signs, including a brief mental test score (see Chapter 2) in order to be able to monitor progress rather than to help in the diagnosis; a careful account of possible sites of infection, including occult infection in the abdomen or other intra-abdominal pathology (for example, the presence of decreased bowel sounds indicative of 'silent' perforation), the presence of constipation or melaena on rectal examination; and careful inspection for ulcers, sores or fractures. The state of the patient's nutrition and hydration is important: patients are usually weighed on admission in hospital but rarely measured, so it is not always possible to relate their weight to an ideal body mass. However, malnutrition sufficient to produce delirium is likely to be apparent without recourse to tables of normal values.

The initial laboratory requests should combine tests to confirm any clinical diagnosis with tests designed to detect common but clinically occult problems (see Table 5.1). While these results are awaited management can and should begin on an empirical basis (see page 80) directed at what is considered to be the most probable cause of the delirium.

In due course, results will be received from the laboratory which will fall into one of three categories. First, there are those that confirm the initial clinical diagnosis of the cause of delirium. Provided the patient is responding to treatment, then all is satisfactory. Second, additional abnormal results will be returned that do not correspond with the initial diagnosis. It is wise to deal with these abnormalities also, whether or not the patient is responding, because they may be contributing to the overall clinical picture. Finally, none of the initial laboratory results may be abnormal. Again, if the patient is responding to empirical treatment, no further investigations need be carried out. If not, or if there is deterioration, then consideration should be given to other possible causes, therapy should be modified in the light of the laboratory and clinical data that are *normal* (thus ruling out a number of possible diagnoses), and more specialized or invasive investigations will need to be undertaken (Table 5.2).

It is as well at this stage (usually two to three days after admission) to

Table 5.1 Initial laboratory investigations in delirium

CLINICALLY SUGGESTED AREA OF INTEREST	TESTS INDICATED (routine tests underlined)
DRUGS	Relevant levels: (phenytoin, digoxin, salicylate). Store serum, urine in case of overdose.
INFECTION	Sputum. Urine (culture, protein, blood). Blood cultures, wound swabs, throat swab. WBC, ESR, (C-reactive protein), chest X-ray. Baseline serum for *Legionella*, viral titres. Stool culture (*Salmonella*, *Clostridium difficile*).
METABOLIC	Electrolytes, urea, creatinine, Ca^{++}, PO_4^{--}, thyroid function, liver function, glucose (urine, blood). Arterial blood gases and pH.
HAEMATOLOGY	Haemoglobin, platelets, vitamin B_{12}, Folate, Sickle cell screen. Carboxy-haemoglobin levels (if history suggestive).
CARDIAC	ECG, CPK, LDH.
ABDOMINAL	Amylase, erect abdominal X-ray.
NEUROLOGICAL	Skull X-ray (if evidence of trauma). Lumbar puncture (if subarachnoid haemorrhage or meningitis suggested clinically).

undertake a full clinical review of the patient, looking particularly for *new* physical signs, or changes in those previously present. At this point, the importance of careful documentation at the time of admission will be apparent. Certain of the initial investigations should also be repeated at this time, particularly those relevant to infection, cardiac failure, and anaemia which may become abnormal over a short space of time.

Treatment of specific causes

Clearly, if a specific cause for the delirium is identified or suspected, it must be treated. Since delirium is most frequently symptomatic of an

Table 5.2 Specialized (second-line) investigations

CLINICALLY SUGGESTED AREA OF INTEREST	TESTS INDICATED (selected according to clinical probability)
DRUGS	As TABLE 5.1 – repeat abnormal levels.
INFECTION	Indium[111] – labelled white cell scan or gallium scan, to identify probable site.
	Echocardiogram for possible vegetations. Repeat initial cultures, WBC, ESR, CRP, chest radiograph. Blood/CSF syphilis serology.
METABOLIC	(Repeat of electrolytes, urea, creatinine, LFTs).
HAEMATOLOGY	Malaria. Bone marrow if film abnormal initially. Repeat full blood picture.
CARDIAC	Echocardiogram if infectious endocarditis or effusion suspected. (repeat ECG, CPK, LDH).

ABDOMINAL	Ultrasound ? Subphrenic or paracolic abscess
	                      ? Neoplasia
	                      ? Other abscess
	Liver biopsy – if indicated.

NEUROLOGICAL	Isotope or CT brain scan (especially if trauma). EEG (also useful to confirm diagnosis). Lumbar puncture if not already carried out.
OTHER	Rarely exacerbation of autoimmune disease, e.g. polymyalgia rheumatica, systemic lupus erythematosus.

underlying organic pathology, this pathology should sought diligently, bearing in mind that there may be multiple abnormalities present. The correction of anaemia, electrolyte imbalance, reduction in sedative medication, treatment of pain, and treatment of infection may all be required in order to resolve the episode speedily. It is not useful to review here the specific treatments for particular medical conditions that may cause delirium; this information is to be found in standard medical texts.

Sometimes the necessary treatment may intensify the delirium in the short term, and it is in the careful balancing of these therapeutic conflicts that expertise is required. For example, the management of delirium caused by pain may require the use of narcotic analgesics. As a rule, the only sources of pain that should be treated initially with narcotic analgesics are fracture and malignancy. In the case of fractures, subsequent immobilization in plaster or an appropriate orthopaedic procedure will allow reduction in analgesic dosage, but in malignancy, continued analgesia may be the only relief available. Early radiotherapy should be considered for localized bone lesions, and non-steroidal anti-inflammatory drugs frequently prove very useful as an adjunct to narcotic analgesics.

Drug withdrawal

Another therapeutic dilemma is posed by cases of delirium caused by drugs. Stopping a drug may initially make the delirium worse, either by a subsequent deterioration of the patient's physical state, or by inducing a hyperactive withdrawal syndrome. In fact, the majority of drugs which cause delirium in the elderly (see Table 3.1) may be withdrawn safely in the hospital setting, since the impact of withdrawal both on the delirium and on the primary disorder for which they were prescribed can be closely monitored. However, the withdrawal of the following groups of drugs requires particular care and attention.

Antiparkinsonian drugs

Severe parkinsonism may be precipitated if all antiparkinsonian treatment is withdrawn suddenly. It is best to withdraw any anticholinergic agents first, and then reduce the dose of dopamine agonist, and perhaps treat the patient with amantadine if the parkinsonism shows signs of worsening.

Anticonvulsants

Blood anticonvulsant levels can usually be obtained to determine the presence of toxicity, and phenytoin for example has a long enough half-life for the dose to be reduced to 100 mg per day without risking recurrence of epilepsy before phenytoin levels are returned from the laboratory. If it is necessary to withdraw the anticonvulsant completely and results may be delayed, then heminevrin or carbamazepine should be introduced to cover the period of drug withdrawal.

Digoxin

Treatment with this drug can usually be stopped while awaiting results from the laboratory. If it has been given for atrial fibrillation, then amiodarone is a good substitute and is well tolerated. However, like digoxin it has a long half-life. It also displaces digoxin from protein binding sites and may therefore temporarily increase free digoxin if it is not introduced gradually.

Sedative drugs

With regard to delirium caused by sedative drugs, it should not be assumed that patients will necessarily become more aggressive if the drugs are withdrawn; often the reverse is true. The exception is withdrawal of drugs of addiction, such as alcohol and opiates, and in these patients adequate sedation is necessary. In view of the increased risk of fits, delirium due to alcohol or drug withdrawal is best managed with drugs that are both sedative and anticonvulsant, such as diazepam, oxazepam, midazolam, or chlormethiazole. These drugs are potentially addictive in their own right, so they should be withdrawn gradually. The simplest and most effective short-term form of sedation in alcohol withdrawal may well be the administration of a little alcohol; this is considerably cheaper and at least as effective as other forms of sedation. Since elderly alcoholics are likely to be nutritionally deficient, any withdrawal regimen must always include a course of a high-potency vitamin preparation containing thiamine to prevent Wernicke's encephalopathy and subsequent Korsakoff's psychosis.

Minimizing iatrogenic discomfort

The exacerbation of delirium by procedures required for investigation and treatment has been referred to. The impact of necessary investigations should be minimized wherever possible; for example, blood for investigations should be taken on one occasion, and not as a series of separate venepunctures. Ideally, invasive investigations should be performed when delirious patients are at their most lucid and co-operative.

Treatment of dehydration may seem to require an intravenous infusion but in an agitated patient this may be difficult and painful, and a less disturbing alternative may be subcutaneous infusion into the anterior abdominal wall through a fine needle. Up to two litres of fluid per day may be given by this route. Feeding may prove difficult, and a nasogastric tube will frequently be pulled out if it is ever successfully sited. Generally, nutritional support can wait 48 hours, but if it is critical, then full parenteral feeding with a central line properly tunnelled subcutaneously using general or local anaesthesia may be the only option.

Bladder catheterization should be avoided if possible, because of continued discomfort, the possibility of inducing septicaemia, and the probability that the catheter will be forcibly removed by the patient, even with the balloon inflated. If catheterization is necessary it may be a worthwhile precaution to take blood cultures ten minutes afterwards to document any bacteraemia that occurs.

On the other hand, incontinence of urine or faeces may be extremely distressing to a previously alert and unimpaired patient. Prompt cleaning with regular toileting may help these problems, as well as providing part of the 'routine' that the delirious patient is said to require (Lipowski 1989).

Psychological aspects of management

Although the relative merits of different treatment regimens in severe agitation due to delirium, especially delirium tremens, have been examined (Salzman 1987; Rosenbloom 1988; Adams 1988; Fernandez *et al.* 1988; Menza *et al.* 1988; Heuzeroth and Gruneklee 1988), there have been no studies formally evaluating the various psychological approaches that are very often suggested as helpful. Most of the advice given by psychiatrists to their colleagues caring for delirious elderly patients on medical and surgical wards is therefore unsupported by any empirical evidence; it is at times self-contradictory; and it often represents a counsel of perfection attainable only in the most luxuriously appointed private hospitals. With these caveats in mind, the advice usually offered is as follows:

(1) *Reduce sensory under-stimulation and maximize the clarity of all perceptions, yet at the same time avoid over-stimulation.*

Quite what the prefixes 'under' and 'over' mean in this context is not clear, since the ideal level and type of sensory stimulation in delirium are not known. Clinical experience indicates that delirious patients should be in a well-lit room without glare or dazzle, and people approaching should be clearly illuminated, stand in full view rather than loom over the bed, and speak slowly, carefully, and distinctly. Holding the patient's hand while talking is an effective means of focusing their attention and providing reassurance. It is likely that the 'ward round', a circus of strange, white-coated, and uniformed figures sweeping in and swirling about the patient, firing questions against a background of muttering comment, is not particularly helpful to anyone.

It is often recommended that delirious patients should be nursed in side rooms in order to minimize the levels of confusing stimulation they are exposed to. In practice, the usual reason for removing delirious patients to side rooms is in order to reduce the disruption to ward routine, and as a result they are often neglected. These patients are usually seriously physically ill, and there is a risk of additional injury from disturbed behaviour and suicide attempts, so they should only be managed in side rooms if there are sufficient staff or relatives to ensure continuous supervision.

Whether or not it is necessary or desirable to keep the patient's room lit at night is debatable. On the one hand, illusions and hallucinations may well be made worse by dark and gloom, but on the other the 24-hour light–dark cycle may be providing important entraining stimuli to the patient's internal clocks.

(2) *Minimize the unfamiliarity of the environment.*

In this category, it is recommended that the patient should have some familiar artefacts near them, one or two relatives and loved ones should be available for visiting, and the same person or persons should consistently nurse the patient.

(3) *Minimize disorientation.*

Place the patient so that they can see a large clock and a marked calendar. Formal Reality Orientation is not applicable to delirium, but repeated reassurances and explanations to the patient by staff are probably helpful.

An often neglected aspect of the psychotherapy of delirium at all ages is support and guidance for the patient's family. They are a vital nursing resource, particularly if the patient is being managed at home, and it is essential that they are fully informed of the causes, treatment, and likely prognosis of the delirium. In particular, they will be worried that the patient has 'gone mad', and will need to be reassured that the delirium is only a temporary affliction. If the patient is demented or physically disabled and has been heavily dependent on family members for care for some time, an episode of delirium may precipitate important deliberations about the viability of further informal care, particularly if the patient has been admitted to hospital. The family's experience of the episode of delirium in terms of stress, workload, support, and outcome will be an important factor in the outcome of these deliberations.

Junior doctors and nursing staff on medical and surgical wards will also benefit from information and reassurance about delirium. Inexperienced staff may be frightened and upset by disturbed, incoherent, and unpredictable patients, and without support and information may be unwilling to devote to them the level of care and supervision that they require. Education based upon particular cases is probably the most effective means of training staff in the detection and management of delirium.

Treatment

Treating the effects of delirium

In the past, physical restraint was recommended as a means of managing disturbed behaviour in delirium; for example Barrough's instruction (1583) that: 'if he be riche let servauntes hold him, if poore, bind him, for inordinate moving diminisheth strength . . .'. This practice continues today, despite the fact that studies of medical, surgical, and institutional samples have repeatedly shown that it is both distressing and potentially dangerous. The employment of physical restraint is usually a nursing decision, and is justified on the grounds that there is no alternative (Strumpf and Evans 1988). Very often this is not in fact the case, and better training of nurses and care staff is clearly needed to change attitudes towards troublesome behaviour, and to enlarge their repertoire of alternative management strategies. In a few cases, physical restraint may be a necessary last resort; if so, its use should be carefully supervised and recorded.

Nowadays, powerful psychotropic drugs are the mainstay of management in psychologically and educationally impoverished hospital and institutional settings where much of the disturbed behaviour due to delirium occurs. However, just as physical restraint compounds the problem by greatly increasing fear and paranoid ideation, so inappropriate sedation can make matters worse by deepening and prolonging the delirium. In the symptomatic management of delirium with sedative drugs a rational prescribing strategy is important to ensure that the benefits always outweigh the disadvantages. Ideally, the cause of the delirium should be known before sedating the patient, since certain drugs may be positively dangerous. For example, a patient who is delirious as a result of respiratory failure may be killed by respiratory depressant sedatives such as benzodiazepines. Where the cause of the delirium is not known, drugs with significant respiratory and cardiac effects should be avoided.

In most cases the disturbed behaviour and distress in delirium are caused by hallucinations and delusions, and a short course of haloperidol is the drug of first choice for the control of these symptoms. It combines a powerful antipsychotic action with relatively low sedative and anticholinergic effects, it is safe in respiratory and cardiac illness, it does not cause excessive postural hypotension, and it can be administered orally, intramuscularly, or intravenously as necessary (Steinhart 1983; Lipowski 1983; Fernandez et al. 1988). Unfortunately, haloperidol is liable to cause acute extrapyramidal side-effects such as dystonia, akathisia, and parkinsonism, and these may pose intolerable problems of their own in some cases. In particular, akathisia (the compulsion to move about) may aggravate rather than relieve any behaviour disturbance associated with delirium. Tardive dyskinesia is an irreversible movement disorder that is typically associated with prolonged neuroleptic treatment, but it can occasionally appear after only a few days of treatment. Extrapyramidal symptoms are less of a problem with other neuroleptics such as chlorpromazine, trifluoperazine and thioridazine, but these compounds are more sedative, more anticholinergic, and may cause cardiac arrhythmias, orthostatic hypotension, and epileptic fits. Thioridazine is probably the most troublesome drug in the elderly, since it has a marked tendency to accumulate. Of the phenothiazines, promazine is associated with the fewest side-effects. Its antipsychotic effects are minimal, but it is useful as a mild, non-addictive sedative.

If neuroleptics are unhelpful or contra-indicated, then benzodiazepines should be considered. The short-term use of compounds with short half-lives, such as oxazepam and temazepam, can provide effective relief of agitation and insomnia, but they will not relieve psychotic symptoms. It is important to bear in mind that benzodiazepines sometimes cause paradoxical excitement and rage. Combinations of sedative drugs are usually discouraged in the elderly, but recent evidence indicates that intravenous haloperidol together with a benzodiazepine may be useful in severe cases of delirium (Adams 1988; Menza *et al.* 1988). There is no longer any place for opiates, barbiturates, or paraldehyde in the management of delirium in the elderly.

There is a wide variation in plasma levels following administration of any given quantity of drug in the elderly (see page 44), so recommended doses are not helpful in the individual case. Since it is difficult to judge the optimum dose in the first instance, the best strategy is to start with a small dose (one quarter that recommended for general adult use) and increase it as necessary. The emergency high-dose 'knock-out' sedative may abolish disturbed behaviour in the short term, but it will prolong the delirium and encourage the vicious circle of further disturbance and sedation. Sedative drugs should be prescribed as regular medication over short periods (24–48 hours) to ensure regular review of the dosage. Regular review of sedative medication is particularly important in patients who are being managed at home.

Treating the causes of delirium

If neurochemically distinct subtypes of delirium do exist and can be identified (see Chapter 1), it may eventually be possible to manage cases with specific treatments, rather than with blanket restraints. The cholinergic hypothesis of delirium is a promising point of departure, and pharmacological agents such as physostigmine are effective in reversing some of the symptoms of delirium, particularly those specifically due to anticholinergic drugs (Jenike 1982) (see page 68). However, this effect is brief, and the drug is potentially dangerous for elderly people with cardiovascular disease (Milam and Bennet 1987). A great deal of research and development will be necessary before specific, effective, and safe antidelirious drugs for the elderly appear on the market.

There have been several reports recommending ECT as an effective treatment for delirium (Roth and Rosie 1953; Roberts 1963; Kramp and Bolwig 1981). In some cases, this response to ECT may be due to the treatment of an affective disorder underlying or mimicking the delirium; as Dubovsky (1986) has pointed out, organic cerebral disorders such as delirium are not a contra-indication to ECT in depressed patients. The effectiveness of ECT in the treatment of delirium itself has yet to be demonstrated. Most courses of ECT last two weeks or more, and until controlled trials are carried out, it remains possible that any 'response' is simply due to spontaneous remission of the delirium during the period of treatment.

Planning discharge and preventing recurrence.

It is important to remember that delirium may be slow to resolve in the elderly, particularly if it is associated with other brain disorders such as dementia or stroke. In the sample described by Koponen et al. (1989b), the mean duration of the delirious episode was 20 days, with a range of 3–81 days. Recovery is usually gradual, with episodes of nocturnal delirium persisting after apparent recovery during the daytime.

In order for discharge to be successful it is important to be certain that relatives and carers understand the acute nature of the disorder, and genuinely believe that the patient has returned nearly to their former state. If relapse is thought likely, relatives and other contacts such as home helps should be informed of this possibility, and told what to look for.

To avoid any recurrence of delirium on discharge home a preliminary home visit and a trial at home may be useful. Occasionally, patients remain relentlessly noisy and disturbed on the ward despite treatment for all obvious causes of delirium, and yet settle down to their previous amenable selves when returned to their familiar home environment. Additional service input may be needed, and of course if there is anything insanitary or unsafe about the patient's home or lifestyle then this should be corrected where possible. Examples include poor nutrition, practical problems with personal hygiene, perhaps due to poor mobility, hoarding of drugs previously prescribed and confusion over dosages, inability of carers to gain access, lack of a telephone, inadequate heating and lighting, and so forth. Atkinson *et al.* (1978) have identified the following strategies for improving compliance with medication:

(1) ensure that the patient understands their drug regime before discharge from hospital;

(2) minimize confusion over drug-taking by consistent prescribing of the same brand of each drug;

(3) take the patient's preferences with regard to taste, size, colour of preparation into consideration where possible;

(4) ensure that containers have caps that can be easily removed.

Some form of follow-up to ensure compliance with drug regimens is also useful.

Those who have indwelling catheters may benefit from continuous prophylactic antibiotics (trimethoprim or perhaps oral ciprofloxacin if resistant organisms are a problem). Patients with anaemia should be fully investigated for the cause of this after the resolution of their delirium. Patients with metabolic disorders, especially diabetes, should have follow-up to ensure continued control, and patients with respiratory or cardiac failure should have follow-up to ensure that the medication that proved adequate in hospital remains so at home.

6 The future: an agenda for research

Delirium presents a considerable challenge to the health services in developed societies as the numbers of elderly people, particularly the very elderly, increase over the next two decades. It complicates many illnesses and procedures, and is associated with high rates of mortality, considerable distress, and prolonged hospital stay. However, despite its evident importance in terms of patient welfare and health economics, little is known about the prevalence and pathogenesis of delirium, and effective means of detection, management, and prevention have yet to be developed and evaluated.

This research agenda is not drawn up simply for the benefit of academics and funding bodies. Although answering some of these questions will require a considerable investment of expertise and money, many others could be addressed by interested clinicians in the course of their daily practice. In any event, since delirium is most likely to manifest in clinical settings, any investigation will always require the co-operation of the involved medical and nursing staff.

Definition

As the other chapters of this book have made clear, the absence of an agreed definition of delirium seriously limits the value of the existing evidence. Studies employing modern case criteria such as DSM-III or ICD10 are broadly comparable, but these definitions of delirium require an organic cause to be present or the outcome to be known, so they cannot be used in aetiological research or in studies of patients while they are ill. Future research into delirium will be greatly assisted if an agreed definition offering the minimum of restrictions is adopted. It is proposed that delirium should be defined strictly as a syndrome with the following features:

(1) abrupt onset;

(2) cognitive decrement from pre-morbid levels of function, particularly in alertness, orientation, attention, and memory;

(3) perceptual disturbances;

(4) persecutory ideas;

(5) affective disturbances;

(6) increased or decreased psychomotor activity;

(7) disturbance of the sleep–wake cycle;

(8) fluctuation of clinical features over the course of a day.

The advantages of such a definition are that it is possible to establish whether or not it applies while the patient is ill; it makes no stipulation about cause, and it is readily transformed into operational terms. The ICD10 criteria (Table 1.2) without the outcome requirement, and the DSM-III-R criteria (Table 1.4) without criterion E both more or less meet this definition.

Assessment

An agreed definition of delirium will not of itself ensure comparability between future studies; it will also be necessary to develop a uniform method of clinical assessment. What is required is an instrument with the following properties:

(1) the ability to distinguish between delirium and other conditions, notably dementia. This will probably entail the serial administration of tests to the patient, together with repeated formal information-gathering from relatives and nursing staff. A wider range of cognitive function items than is currently available in brief tests such as the MMSE will probably be needed if this instrument is to be a sensitive and specific identifier of delirium.

(2) The ability to estimate reliably changes in the severity of delirium over time. This will be neccesary if the efficacy of drugs and other procedures in the management of delirium are to be evaluated.

It will also be necessary to develop and evaluate procedures for screening vulnerable populations, particularly medical and surgical in-patients.

Epidemiological studies

Armed with an operationalized definition of delirium and a satisfactory means of detection, it will then be possible to determine the prevalence, outcome, and aetiology of delirium in the elderly in various settings. To date, most studies have been of in-patient populations; little or nothing is known about the prevalence of delirium in continuing care facilities, in primary care populations, or in the community. Elderly people in nursing homes and residential care are a particularly vulnerable group so far as the risk of delirium is concerned. They have high rates of chronic physical and mental disorder (Mann *et al.* 1984; Rovner *et al.* 1986), and they receive large quantities of psychotropic medication, often inappropriately (Gurland *et al.* 1979; Mann *et al.* 1986).

Virtually nothing is known about the phenomenology, presentation, detection or aetiology of delirium in elderly people belonging to ethnic minorities in the UK. To what extent do language difficulties interfere with the diagnostic process? Is there any evidence of racist over- or under-diagnosis of delirium in these groups? Are there any particular cultural variables that predispose to delirium in the various communities?

Further studies of the outcome of delirium in the elderly are needed in order to determine the personal, social, and financial costs of this condition. Is the increased mortality associated with delirium due entirely to underlying physical illness, or does delirium itself worsen the prognosis? What is the effect of an episode of delirium upon the long-term provision of care to a physically or mentally disabled elderly individual? For example, does it increase or decrease the likelihood of subsequent institutionalization? Do different clinical types of delirium differ in outcome in old age? Delirium has been shown to be associated with increased length of hospital stay across a wide range of diagnostic groups (Thomas *et al.* 1988): to what extent is prolonged stay due specifically to the delirium – for example, as a result of delays caused by the investigation and diagnosis of the abnormal mental state?

Some clarification of the relative importance of risk factors such as physical illness, atheroma, dementia, functional psychiatric disorder, alcohol abuse, and sensory impairment is still needed. Are these independent of each other, or are they related via other variables, such as age or socio-economic group? Specific aetiological factors also need to be identified, and case-control studies comparing delirious patients in various target populations with appropriately matched controls will be helpful in this respect, as will prospective follow-up studies of particular groups, such as patients undergoing elective surgery, consumers of tricyclic antidepressants, or individuals due for relocation. It is usually assumed that a past history of delirium is predictive of future episodes, but this has yet to be demonstrated. The identification of factors that are either aetiological, or else pose a significant independent risk of delirium will be an important part of the development of preventive strategies.

Pathogenesis

Delirious patients are usually physically ill, unco-operative, and not competent either to give or to withhold consent to invasive procedures, and as a result the neuronal basis of delirium is very difficult to study in humans. Most of the neurophysiological and neuropharmacological evidence about the function of neuronal systems thought to be involved in delirium is derived from animal studies, and its relevance is limited accordingly. Future studies of the pathogenesis of delirium in humans will require either the participation of informed volunteers (elective surgical patients?) in invasive procedures, or else the development of non-invasive techniques that can be ethically applied to non-competent patients. Functional imaging techniques are unlikely to provide much helpful information about delirium in their present form, because of the level of co-operation required from the patient. Procedures such as somatosensory evoked potentials that require little or no co-operation provide more promising avenues for exploration at present.

Study of the pathogenetic mechanisms underlying specific causes of delirium may lead to the development of intervention strategies. For example, what is it about infections that can cause delirium – decrease in cerebral oxygen supply; decrease in blood pressure; 'septicaemia'; or the release of endogenous pyrogens such as interleukin-1? More generally, it is likely that a proper understanding of the pathogenesis of delirium will tell us a great deal about the normal functioning of the brain.

Evaluation of interventions

With appropriate methods, it will be possible to evaluate the effectiveness of various treatment, management, and preventive strategies. Many questions await answers: does pre-operative counselling and reassurance significantly reduce the rate of post-operative delirium, as suggested by Smith and Dimsdale (1989)? Does the delirious in-patient actually benefit from exposure to familiar faces / artefacts / continuous illumination, as is usually recommended? Does sensory deprivation precipitate delirium? What is the optimal level of stimulation? Which management strategies are practicable on a busy ward in a general hospital? Which aspects of normal ward routine are positively anti-therapeutic for the elderly delirious patient? Is haloperidol the most effective sedative in delirium? Do reversible anticholinesterases such as physostigmine have a role in the treatment of delirium? What about other treatment strategies, such as ECT, or 'arousing' medication as in certain hyperactive children? In which cases is management of delirium at home feasible, and when is admission to hospital indicated? Does rapid, intensive treatment of the cause of the delirium hasten resolution or not? What is the role of antibiotic prophylaxis in preventing delirium in elderly people with recurrent chest and urinary tract infections?

The supposed all-or-nothing outcome characteristics of delirium have been a serious obstacle to research. If delirium either heralds death, or else is simply a nuisance on the road to recovery, then there is little incentive to seeking a detailed understanding of how it is caused and how it might be best managed or prevented. However, there are signs that this state of affairs is changing, and that delirium in the elderly is beginning to attract the attention of epidemiologists, neuroscientists, and clinical researchers. It is to be hoped that their efforts will enhance the understanding and care of delirium at all ages.

References

Adams, F. (1988). Emergency intravenous sedation of the delirious, medically ill patient. *Journal of Clinical Psychiatry*, **49**, Supp., 22–7.

Adams, T. (1861). *The Works*. Vol. 1, p. 254.

Aggernaes, A. and Myschetzky, A. (1976). Experienced reality in somatic patients more than 65 years old. *Acta Psychiatrica Scandinavica*, **54**, 225–37.

Alimkhanov, Z. A. (1988). The results of population-based psychopathologic research of paranoid schizophrenia. Translated abstract. *Zhournal Nevropatologii Psikhiatrii*, **88**, 70–6.

American Psychiatric Association (1980). *Diagnostic and statistical manual of mental disorders, (3rd edn)*. American Psychiatric Association, Washington, DC.

American Psychiatric Association (1987). *Diagnostic and statistical manual of mental disorders*, (3rd edn, revised). American Psychiatric Association, Washington, DC.

Anthony, J. C., LeResche, L., Niaz, U., Von Korff, M. R., and Folstein, M. F. (1982). Limits of the 'Mini-Mental State' as a screening test for dementia and delirium among hospital patients. *Psychological Medicine*, **12**, 397–408.

Atkinson, L., Gibson, I., and Andrews, J. (1978). An investigation into the ability of elderly patients continuing to take drugs after discharge from hospital, and recommendations concerning improving the situation. *Gerontology*, **24**, 225–34.

Bachelard, H. (1975). Energy utilized by neurotransmitters. In *Brain work: proceedings of the Alfred Benzon symposium VIII*, (ed. D. H. Ingvar and N. A. Lassen) pp. 79–81. Munksgaard, Copenhagen.

Balter, R. A., Fricchione, G., and Sterman, A. B. (1986). Clinical presentation of multi-infarct delirium. *Psychosomatics*, **27**, 461–2.

Banks, C. M. and Vojnik, M. (1978). Comparative simultaneous measurement of cerebrospinal fluid 5-hydroxyindoleacetic acid and blood serotonin levels in delirium tremens and clozapine-induced delirious reaction. *Journal of Neurology, Neurosurgery and Psychiatry*, **41**, 420–4.

Baraldi, M. and Zeneroli, M. L. (1982). Experimental hepatic encephalopathy: Changes in the binding of gamma-aminobutyric acid. *Science*, **216**, 427–9.

Barrough, P. (1583). *The Methode of Phisicke, conteyning the causes, signs, and cures of inward diseases in mans body from the head to the foote*. Vautrollier, London.

Bartholomeus Anglicus (1535). *De proprietatibus rerum*. Barthelet, London.

Basavaraju, N. and Phillips, S. L. (1989). Cortisol deficient state. A case of reversible cognitive impairment and delirium in the elderly. *Journal of the American Geriatrics Society*, **37**, 49–51.

Bedford, P. D. (1959). General medical aspects of confusional states in elderly people. *British Medical Journal*, **ii**, 185–8.

Bell, R. and Brown, K. (1985). Cholinergic mechanisms in aggressive behavior: Role of muscarinic and nicotinic systems. In *Central cholinergic mechanisms and adaptive dysfunctions* (ed. M. M. Singh, D. M. Warburton and H. Lal) pp. 161–192. Plenum Press, New York.

Beresin, E. (1988). Delirium in the elderly. *Journal of Geriatric Psychiatry and Neurology*, **1**, 127–43.

Berggren, D., Gustafson, Y., Eriksson, B., *et al.* (1987). Post-operative confusion after anesthesia in elderly patients with femoral neck fractures. *Anesthesia and Analgesia*, **66**, 497–504.

Bergmann, K. and Eastham, E. J. (1974). Psychogeriatric ascertainment and assessment for treatment in an acute medical setting. *Age and Ageing*, **3**, 174–88.

Berman, P. and Fox, R. A. (1985). Fever in the elderly. *Age and Ageing*, **14**, 327–32.

Berrios, G. E. (1981). Delirium and confusion in the 19th Century: a conceptual history. *British Journal of Psychiatry*, **139**, 439–49.

Berrios, G. E. and Brook P. (1982). The Charles Bonnet syndrome and the problem of visual perceptual disorders in the elderly. *Age and Ageing*, **11**, 17–23.

Berry, D. T. R., Philips, B. A., Cook, Y. R., Schmitt, F. A., Gilmore, R. L., Patel, R., Keener, T. M., and Tyre, E. (1987). Sleep-disordered breathing in healthy aged persons: possible daytime sequelae. *Journal of Gerontology*, **42**, 620–6.

Bickford, R. G. and Butt, H. R. (1955). Hepatic coma: the EEG pattern. *Journal of Clinical Investigation*, **34**, 790–9.

Bienenfeld, D. and Wheeler, B. G. (1989). Psychiatric services to nursing homes: A liaison model. *Hospital and Community Psychiatry*, **40**, 793–4.

Blass, J. P. and Plum, F. (1983). Metabolic encephalopathies. In *The neurology of aging* (ed. R. Katzman and R. D. Terry). F. A. Davis, Philadelphia.

Blazer, D. G., Federspiel, C. F., Ray, W. A. *et al.* (1983). The risk of anti-cholinergic toxicity in the elderly: A study of prescribing practices in two populations. *Journal of Gerontology*, **38**, 31–5.

Blessed, G., Tomlinson, B. E., Roth, M. (1968). The association between quantitative measures of dementia and of senile change in the cerebral grey matter of elderly subjects. *British Journal of Psychiatry*, **114**, 797–811.

Bleuler, M., Willi, J., and Buhler, H. R. (1966). *Akute Psychische Begleiterscheinungen Körperlicher Krankheiten*. Thieme Verlag, Stuttgart.

Blundell, E. (1967). A psychological study of the effects of surgery on eighty-six elderly patients. *British Journal of Social and Clinical Psychology*, **6**, 297–303.

Bond, J. (1987). Psychiatric illness in later life. A study of prevalence in a Scottish population. *International Journal of Geriatric Psychiatry*, **2**, 39–58.

Bond, T. C. (1980). Recognition of acute delirious mania. *Archives of General Psychiatry*, **37**, 553–4.

Bonhoeffer, K. (1910). *Symptomatic psychoses*. Springer, Berlin.

Borchardt, C. M., and Popkin, M. K. (1987). Delirium and the resolution of depression. *Journal of Clinical Psychiatry*, **48**, 373–5.

Braithwaite, R. (1982). The pharmacokinetics of psychotropic drugs in the elderly. In *Psychopharmacology of Old Age* (ed. D. Wheatley) pp. 46–54. Oxford University Press, Oxford.

Cameron, D. J., Thomas, R. I., Mulvihill, M., and Bronheim, H. (1987). Delirium: A test of the Diagnostic and Statistical Manual III criteria on medical inpatients. *Journal of the American Geriatrics Society*, **35**, 1007–10.

Cammarata, R. J., Rodnam, G. P. and Fennell, R. H. (1967). Serum antigamma-globulin and anti-nuclear factors in the aged. *Journal of the American Medical Association*, **199**, 115–118.

Cantrell, M. and Yoshikawa, T. T. (1984). Infective endocarditis in the aging patient. *Gerontology*, **30**, 316–26.

Caplan, L. R., Kelly, M., Kase, C. S. *et al.* (1986). Mirror image of Wernicke's aphasia. *Neurology*, **36**, 1015–20.

Catalan, J., Gath, D. H., Bond, A. *et al.* (1988). General practice patients on long-term psychotropic drugs. A controlled investigation. *British Journal of Psychiatry*, **152**, 399–405.

Chandarana, P. C., Cooper, A. J., Goldbach, M. M. *et al.* (1988). Perceptual and cognitive deficit following coronary artery bypass surgery. *Stress Medicine*, **4**, 163–71.

Chedru, F. and Geschwind, N. (1972*a*). Disorders of higher cortical functions in acute confusional states. *Cortex*, **8**, 395–411.

Chedru, F. and Geschwind, N. (1972*b*). Writing disturbances in acute confusional states. *Neuropsychologia*, **10**, 343–53.

Christoloudou, G. N. (1977). The syndrome of Capgras. *British Journal of Psychiatry*, **130**, 556–64.

Clark, A. N. G., Mankikar, G. D., and Gray, I. (1975). Diogenes syndrome. A clinical study of gross neglect in old age. *Lancet*, **i**, 366–8.

Coben L. A., Danziger, W. L. and Hughes, C. P. (1983). Visual evoked potentials in mild senile dementia of the Alzheimer type. *Electroencephalography and Clinical Neurophysiology*, **55**, 121–30.

Crossland, J., Elliott, K. A., and Pappius, H. M. (1955). Acetylcholine content of brain during insulin hypoglycaemia. *American Journal of Physiology*, **183**, 23–35.

Cutting, J. (1980). Physical illness and psychosis. *British Journal of Psychiatry*, **136**, 109–19.

Cutting, J. (1987). The phenomenology of acute organic psychosis. Comparison with acute schizophrenia. *British Journal of Psychiatry*, **151**, 324–32.

Damas-Mora, J., Skelton-Robinson, A. and Jenner, F. A. (1982). The Charles Bonnet syndrome in perspective. *Psychological Medicine*, **12**, 251–61.

Danielczyk, W. (1984). Pharmacotoxic psychosis in patients with neurological disorders in old age. *Advances in Neurology*, **40**, 285–8.

Davison, K. (1989). Acute organic brain syndromes. *British Journal of Hospital Medicine*, **41**, 89–92.

Devinsky, O., Bear, D. and Volpe, B. T. (1988). Confusional states following posterior cerebral artery infarction. *Archives of Neurology*, **45**, 160–3.

DHSS (1976) The management of violent and potentially violent patients in hospital. *Circular HC* (76) 11.

Dick, J. P. R., Guillof, R. J., Stewart, A. *et al.* (1984). Mini-Mental State examination in neurological patients. *Journal of Neurology, Neurosurgery and Psychiatry*, **47**, 496–9.

Dilsaver, S. C., Snider, R. M. and Alessi, N. E. (1987). Amitriptyline supersensitizes a central cholinergic mechanism. *Biological Psychiatry*, **22**, 495–505.

Domino, E. F., Krause, R. R. and Bowers, J. (1973). Regional distribution of some enzymes involved with putative neurotransmitters in the human visual system. *Brain Research*, **58**, 79–189.

Dongier, M. (1974). Clinical EEG III. In *Handbook of EEG and clinical neurophysiology* (ed. A. Remond). Elsevier, Amsterdam.

Doty, E. J. (1946). The incidence and treatment of delirious reactions in later life. *Geriatrics*, **1**, 21–26.

Downton, J. H., Andrews, K., and Puxtey, J. A. H. (1987). 'Silent' pyrexia in the elderly. *Age and Ageing*, **16**, 41–4.

Drinka, P. J., Voeks, S. K., Langer, E. H. (1988). The psychiatric symptoms of Alzheimer's disease. *Journal of the American Geriatrics Society*, **36**, 961–2.

Drummond, G. B. (1975). The assessment of postoperative mental function. *British Journal of Anaesthesia*, **47**, 130–42.

Dubovsky, S. L. (1986). Using electroconvulsive therapy for patients with neurological disease. *Hospital and Community Psychiatry*, **37**, 819–25.

Dunne, J. W., Leedman, P. J., and Edis, R. H. (1986). Inobvious stroke: a cause of delirium and dementia. *Australian and New Zealand Journal of Medicine*, **16**, 771–8.

Duvoisin, R. C. and Katz, R. D. (1968). Reversal of central anticholinergic syndrome in man by physostigmine. *Journal of the American Medical Association*, **290**, 1963–5.

Easton, C. and MacKenzie, F. (1988). Sensory-perceptual alterations: delirium in the intensive care unit. *Heart and Lung*, **17**, 229–37.

Edwards, C. R. (1977). Vasopressin and oxytocin in health and disease. *Clinics in Endocrinology and Metabolism*, **6**, 223–59.

Engel, G. L. and Romano, J. (1959). Delirium, a syndrome of cerebral insufficiency. *Journal of Chronic Diseases*, **4**, 260–76.

Erkinjuntti, T., Wikstrom, J., Palo, J., and Autio, L. (1986). Dementia among medical inpatients. Evaluation of 2000 consecutive admissions. *Archives of Internal Medicine*, **146**, 1923–6.

Erkinjuntti, T., Sulkava, R., Wikstrom, J., and Autio, L. (1987). Short Portable Mental Status Questionnaire as a screening test for dementia and delirium among the elderly. *Journal of the American Geriatrics Society*, **35**, 412–16

Evans, L. K. (1987). Sundown syndrome in institutionalized elderly. *Journal of the American Geriatrics Society*, **35**, 101–8.

Faust, D., and Fogel, B. S. (1989). The development and initial validation of a sensitive bedside cognitive screening test. *Journal of Nervous and Mental Disease*, **177**, 25–31.

Fernandez, F., Holmes, V. F., Adams, F., and Kavanaugh, J. J. (1988). Treatment of severe, refractory agitation with a haloperidol drip. *Journal of Clinical Psychiatry*, **49**, 239–41.

Fink, M., Green, M., and Bender, M. B. (1952). The face-hand test as a diagnostic sign of organic mental syndrome. *Neurology*, **2**, 46–58.

Fischer, J. E. and Baldessarini, R. J. (1971). False neurotransmitters and hepatic failure. *Lancet*, **ii**, 75–80.

Fish, F. (1974). *Clinical Psychopathology* (ed. M. Hamilton). Wright, Bristol.

Folstein, M. F., Folstein, S. E., McHugh, P. R. *et al.* (1975). Mini-mental state – a practical method for grading the cognitive state of patients for the clinician. *Journal of Psychiatric Research*, **12**, 189–98.

Fontaine, R., Chouinard, G., and Annable, L. (1984). Rebound anxiety in anxious patients after abrupt withdrawal of benzodiazepine treatment. *American Journal of Psychiatry*, **141**, 848–52.

Foy, A., Drinkwater,V., March, S., and Mearrick, P. (1986). Confusion after admission to hospital in elderly patients using benzodiazepines. *British Medical Journal*, **293**, 1072.

Fraser, R. M. and Glass, I. B. (1978). Recovery from ECT in elderly patients. *British Journal of Psychiatry*, **133**, 524–8.

Fraser, R. M. and Glass, I. B. (1980). Unilateral and bilateral ECT in elderly patients. *Acta Psychiatrica Scandinavica*, **62**, 13–31.

Friedman, J. H. (1985). Syndrome of diffuse encephalopathy due to non-dominant thalamic infarction. *Neurology*, **35**, 1524–6.

Fujimoto, A., Nagao, T., Ebora, T. *et al.* (1983) Cerebrospinal fluid monoamine metabolites during alcohol withdrawal syndromes and recovered state. *Biological Psychiatry*, **18**, 1141–52.

Furstenberg, A. L. and Mezey, M. D. (1987). Mental impairment of elderly hospitalised hip fracture patients. *Comprehensive Gerontology*, **1**, 80–5.

Garrison, F. H. (1929). *An introduction to the history of medicine*. W. B. Saunders, Philadelphia.

Geschwind, N. (1982). Disorders of attention: A frontier in neuropsychology. *Philosophical Transactions of the Royal Society of London*, **298**, 173–85.

Gloor, P., Ball, G., and Schaul, N. (1977). Brain lesions that produce delta waves in the EEG. *Neurology*, **27**, 326–33.

Goffman, E. (1968). *Asylums: essays on the social situation of mental patients and other inmates*. Penguin, Harmondsworth.

Gold, K. and Rabins, P. V. (1989). Isolated visual hallucinations and the Charles Bonnet syndrome: A review of the literature and six cases. *Comprehensive Psychiatry*, **30**, 90–8.

Goldberg, D., Cooper, B., Eastwood, M., Kedward, H., and Shepherd, M. (1970). A psychiatric interview suitable for using in community surveys. *British Journal of the Society for Preventive Medicine*, **24**, 18–26.

Golinger, R. C. (1986). Delirium in surgical patients seen at psychiatric consultation. *Surgery, Gynecology and Obstetrics*, **163**, 104–6.

Granacher, R. P. and Baldessarini, R. J. (1975). Physostigmine. Its use in acute anticholinergic syndrome with antidepressant and antiparkinsonian drugs. *Archives of General Psychiatry*, **32**, 375–80.

Green, R. L., McAllister, T. W. and Bernat, J. L. (1987). A study of crying in medically and surgically hospitalized patients. *American Journal of Psychiatry*, **144**, 442–7.

Greene, L. T. (1971). Physostigmine treatment of anticholinergic delirium in post-operative patients. *Anesthesia and Analgesia*, **50**, 222–6.

Grimley Evans, J. (1982). The psychiatric aspects of physical disease. In *The psychiatry of late life* (ed. R. Levy and F. Post). pp. 114–142. Blackwell Scientific Publications, Oxford.

Gurland, B. J., Kuriansky, J. B., Sharpe, L., Simon R., Stiller, P., and Birkett, P. (1977). The Comprehensive Assessment and Referral Examination (CARE) – rationale, development and reliability. *International Journal of Ageing and Human Development*, **8**, 9–42.

Gurland, B. J., Cross, P., Defiguerido, J. *et al.* (1979). A cross-national comparison of institutionalised elderly in the cities of New York and London. *Psychological Medicine*, **9**, 781–8.

Gustafson, Y., Berggren, D., Brannstrom, B. *et al.* (1988). Acute confusional states in elderly patients treated for femoral neck fracture. *Journal of the American Geriatrics Society*, **36**, 525–30.

Haefely, W. (1986). Biological basis of drug-induced tolerance, rebound and dependence. Contribution of recent research on benzodiazepines. *Pharmacopsychiatry*, **19**, 353.

Hartley, D. (1749). *Observations on man, his frame, his duty, and his expectations*. Leake and Frederick, London.

Hawley, R. J., Major, L. F., Schulman, E. A. *et al.* (1981). CSF levels of norepinephrine during alcohol withdrawal. *Archives of Neurology*, **38**, 289–92.

Hector, M. and Burton, J. R. (1988). What are the psychiatric manifestations of vitamin B_{12} deficiency? *Journal of the American Geriatrics Society*, **36**, 1105–12.

Heinrich, C. P., Stadtler, H., and Weiser, H. (1973). The effect of thiamine deficiency on the acetylcoenzyme A and acetyl choline levels in the rat brain. *Journal of Neurochemistry*, **21**, 1273–81.

Heller, S. S., Frank, K. A., Malm, J. R., Bowman, F. O., Harris, P. D., Charlton, M. H., and Kornfeld, D. S. (1970). Psychiatric complications of open-heart surgery: A re-examination. *New England Journal of Medicine*, **283**, 1015–20.

Heller, S. S., Kornfeld, D. S., Frank, K. A., and Hoar, P. F. (1979). Postcardiotomy delirium and cardiac output. *American Journal of Psychiatry*, **136**, 337–9.

Hemmingsen, R., Vorstrup, S., Clemmesen, L. *et al.* (1988). Cerebral blood flow during delirium tremens and related states studied with Xenon-133 inhalation tomography. *American Journal of Psychiatry*, **145**, 1384–90.

Henderson, A. S. (1988). The risk factors for Alzheimer's disease: a review and a hypothesis. *Acta Psychiatrica Scandinavica*, **78**, 257–75.

Henderson, D. K. and Gillespie, R. D. (1936). *A textbook of psychiatry*, (4th edn). Oxford University Press, Oxford.

Henry, G. W. (1935). *Essentials of psychopathology*. Wood and Company, Baltimore.

Hernandez-Peon, R., Chavez-Ibarra, G., Morgane, P. J. *et al.* (1963). Limbic cholinergic pathways involved in sleep and emotional behaviour. *Experimental Neurology*, **8**, 93–111.

Hess, W. R. (1969). *Hypothalamus and thalamus: experimental documentation*. George Thieme, Stuttgart.

Heuzeroth, L. and Gruneklee, D. (1988). Clonidine – an alternative in the treatment of delirium tremens. *Medizinische Klinik*, **25**, 783–9.

Hill, J. D., Hampton, J. R., and Mitchell, J. R. A. (1979). Home or hospital for myocardial infarction – who cares? *American Heart Journal*, **98**, 545–7.

Hodding, G. C., Jann, M., and Ackerman, I. P. (1980). Drug withdrawal syndromes – a literature review. *Western Journal of Medicine*, **133**, 383–91.

Hodkinson, H. M. (1973). Mental impairment in the elderly. *Journal of the Royal College of Physicians*, **7**, 305–17.

Hunter, J. (1835). *The Works of John Hunter, FRS*: Vol. 1 (ed. J. F. Palmer). Longman, London.

Itil, T. and Fink, M. (1966). Anticholinergic drug-induced delirium: experimental modification, quantitative EEG and behavioural correlations. *Journal of Nervous and Mental Disease*, **143**, 492–507.

Jacoby, R. and Bergmann, K. (1986). The psychiatry of old age. In *Essentials of postgraduate psychiatry* (ed. P. Hill, R. Murray and A. Thorley). pp. 495–526. Sure and Shatton, London.

James, J. H., Ziparo, V., and Jeppsson, B. (1979). Hyperammonaemia, plasma amino acid imbalance and blood-brain amino acid transport: A unified theory of portal-systemic encephalopathy. *Lancet*, **ii**, 772–5.

Jasper, H. H. and Tessier, J. (1971). Acetylcholine liberation from the cerebral cortex during paradoxical (REM) sleep. *Science*, **172**, 601–2.

Jelliffe, S. E. and White, W. (1929). *Diseases of the nervous system* (5th edn). Lea and Febiger, Philadelphia.

Jenike, M. A. (1982). Anticholinergic delirium: Diagnosis and treatment. *Topics in Geriatrics*, **1**, 6–7.

Kahn, R. L., Goldfarb, A. I., Pollack, M. *et al.* (1960). Brief objective measures for the determination of mental status in the aged. *American Journal of Psychiatry*, **117**, 326–8.

Kales, A. J., Caldwell, A. B., Cadieux, R. J., Vela-Bueno, A., Ruch, L. G., and Mayes, S. D. (1985). Severe obstructive sleep apnoea. II: associated psychopathology and psychosocial consequences. *Journal of Chronic Diseases*, **38**, 427–34.

Kay, D. W. K., Beamish, P., and Roth, M. (1964). Old age mental disorders in Newcastle-upon-Tyne. Part II: A study of possible social and medical causes. *British Journal of Psychiatry*, **110**, 668–82.

Kellam, A. M. P. (1987). The neuroleptic malignant syndrome, so-called: A survey of the world literature. *British Journal of Psychiatry*, **150**, 752–9.

Kendell, R. E. (1975). The concept of disease and its implications for psychiatry. *British Journal of Psychiatry*, **127**, 305–15.

Kennedy, A. (1959). Psychological factors in confusional states in the elderly. *Gerontology Clinics*, **1**, 71–82.

Khantzian, E. J. and McKenna, G. J. (1979). Acute toxic and withdrawal reactions associated with drug use and abuse. *Annals of Internal Medicine*, **90**, 361–72.

Kiloh, L. G., McComas, A. J., Osselton, J. W., and Upton, A. R. M. (1981). *Clinical electroencephalography* (4th edn) p. 166. Butterworths, London.

Kleinfeld, M., Peter, S., and Gilbert, G. M. (1984). Delirium as the predominant manifestation of hyperparathyroidism: reversal after parathyroidectomy. *Journal of the American Geriatrics Society*, **32**, 689–90.

Knights, E. B. and Folstien, M. F. (1977). Unsuspected emotional and cognitive disturbance in medical patients. *Annals of Internal Medicine*, **87**, 723–4.

Koponen, H., Hurri, L., Stenbäck, U., and Riekkinen, P. J. (1987). Acute confusional states in the elderly: a radiological evaluation. *Acta Psychiatrica Scandinavica*, **76**, 726–31.

Koponen, H., Hurri, L., Stenbäck, U. *et al.* (1989a). Computed tomography findings in delirium. *Journal of Nervous and Mental Disease*, **177**, 226–31.

Koponen, H., Stenbäck, U., Mattila, E. *et al.* (1989b). Delirium in elderly persons admitted to a psychiatric hospital: clinical course during the acute stage and one-year follow-up. *Acta Psychiatrica Scandinavica*, **79**, 579–85.

Koponen, H., Stenback, U., Mattila, E. *et al.* (1989c) CSF beta-endorphin-like immunoreactivity in delirium. *Biological Psychiatry* **25**, 938–44.

Koponen, H., Stenback, U., Mattila, E. *et al.* (1989d) Cerebrospinal fluid somatostatin in delirium. *Psychological Medicine*, **19**, 605–9.

Koponen, H., Partanen, J., Paakkonen, A. *et al.* (1989e) EEG spectral analysis in delirium. *Journal of Neurology and Neurosurgery and Psychiatry*, **52**, 980–5.

Kornfeld, D. S. Heller, S. S. Frank, K. A. and Moskowitz, R. (1974). Personality and psychological factors in post-cardiotomy patients. *Archives of General Psychiatry*, **31**, 249–53.

Kraeplin, E. (1889). *Lehrbuch der Psychiatrie*. Barth, Leipzig.

Kral, V. A. (1975). Confusional states: description and management. In *Modern perspectives in the psychiatry of old age* (ed. J. G. Howells). Brunner/Mazel, New York.

Kramp, P. and Bolwig, T. G. (1981). Electroconvulsive therapy in acute delirious states. *Comprehensive Psychiatry*, **22**, 368–71.

Kräupl-Taylor, F. (1976). The medical model of the disease concept. *British Journal of Psychiatry*, **128**, 588–94.

Krauthammer, C. and Klerman, G. L. (1978). Secondary mania. Manic syndromes associated with antecedent physical illness or drugs. *Archives of General Psychiatry*, **35**, 1333–9.

Krueger, J. M., Waltor, J., Dinarello, C. A., Wolff, S. M. and Chedid, L. (1984). Sleep-promoting effects of endogenous pyrogen (interleukin-1). *American Journal of Physiology*, **246**, 994–9.

Kuhr, B. M. (1979). Prolonged delirium with propanolol. *Journal of Clinical Psychiatry*, **40**, 198–9.

Leechuy, I., Abrams, R., and Kohlhaas, J. (1988). ECT-induced postictal delirium and electrode placement. *American Journal of Psychiatry*, **145**, 880–1.

Levenson, J. L. (1985). Neuroleptic malignant syndrome. *New England Journal of Medicine*, **313**, 163–6.

Levey, A. I., Hallanger, A. E. and Wainer, B. H. (1987). Cholinergic nucleus basalis neurons may influence the cortex via the thalamus. *Neurosciences Letters*, **74**, 7–13.

Levkoff, S. E., Safran, C., Cleary, P. D. *et al.* (1988). Identification of factors associated with the diagnosis of delirium in elderly hospitalized patients. *Journal of the American Geriatrics Society*, **36**, 1099–104.

Levin, M. (1956). Varieties of disorientation. *Journal of Mental Science*, **102**, 619–23.

Liappas, I. A., Jenner, F. A. and Vincente, B. (1987). Withdrawal syndromes. *Journal of the Royal College of Physicians*, **21**, 214–18.

Lindesay, J. (1986). Trends in self-poisoning in the elderly 1974–1983. *International Journal of Geriatric Psychiatry*, **1**, 37–42.

Lindesay, J., Briggs, K., and Murphy, E. (1989). The Guy's / Age Concern Survey: Prevalence rates of cognitive impairment, depression and anxiety in an urban elderly community. *British Journal of Psychiatry*, **155**, 317–29.

Lindesay, J. (1990). The Guy's / Age Concern Survey: Physical health and psychiatric disorder in an urban elderly community. *International Journal of Geriatric Psychiatry*, **5**, 171–78.

Lindsley, J. G. (1983). Sleep patterns and functions. In *Physiological correlates of human behaviour*, Vol. 1 (ed. A. Gale and J. A. Edwards). pp. 105–41. Academic Press, London.

Lipowski, Z. J. (1980*a*). *Delirium: acute brain failure in man.* C. C. Thomas, Springfield, Illinois.

Lipowski, Z. J. (1980*b*). Delirium updated. *Comprehensive Psychiatry*, **21**, 190–6.

Lipowski, Z. J. (1983). Transient cognitive disorders in the elderly. *American Journal of Psychiatry*, **140**, 1426–36.

Lipowski, Z. J. (1985). Delirium (acute confusional state). In *Handbook of clinical neurology* (ed. P. J. Vinken, G. W. Bruyn, and H. L. Klawans) Vol. 46; pp. 523–59. Elsevier Science Publishers, Amsterdam.

Lipowski, Z. J. (1989). Delirium in the elderly patient. *New England Journal of Medicine*, **320**, 578–82.

Lishman, A. (1987). *Organic psychiatry*, (2nd edn). Blackwell Scientific Publications, Oxford.

Loach, A. B. and Benedict, C. R. (1980). Plasma catecholamine concentrations associated with cerebral vasospasm. *Journal of Neurological Sciences*, **45**, 261–71.

Lukianowicz, N. (1958). Autoscopic phenomena. *Archives of Neurology and Neuropsychiatry*, **80**, 199–220.

Lugaresi, E., Medori, R., Montagna, P. *et al.* (1986). Fatal familial insomnia and dysautonomia with selective degeneration of thalamic nuclei. *New England Journal of Medicine*, **315**, 997–1003.

Macdonald, A. J. D., Simpson, A. and Jenkins, D. (1989). Delirium in the elderly: A review and a suggestion for a research programme. *International Journal of Geriatric Psychiatry*, **4**, 311–19.

Mackenzie, T. B. and Popkin, M. K. (1980). Stress response syndrome occuring after delirium. *American Journal of Psychiatry*, **137**, 1433–5.

Maclean, D. and Emslie-Smith, D. (1977). *Accidental hypothermia.* Blackwell Scientific Publications, Oxford.

Macmillan, D. and Shaw, P. (1966). Senile breakdown in standards of personal and environmental cleanliness. *British Medical Journal*, **ii**, 1032–7.

Mandell, G. L. and Sande, M. A. (1985). Antimicrobial agents: penicillins, cephalosporins and other beta-lactam antibiotics. In *Goodman and Gilman's The pharmacological basis of therapeutics* (eds. A. G. Gilman, L. S. Goodman, T. W. Rall, and F. Murad). (7th edn), pp. 1115–49. Macmillan, New York.

Mann, A. H., Graham, N. and Ashby, D. (1984). Psychiatric illness in residential homes for the elderly: a survey in one London borough. *Age and Ageing*, **13**, 257–65.

Mann, A. H., Graham, N., and Ashby, D. (1986). The prescription of psychotropic medication in local authority old people's homes. *International Journal of Geriatric Psychiatry*, **1**, 25–30.

Marsh, G. R. and Thompson, L. V. (1977). Psychophysiology of aging. In *Handbook of the psychology of aging* (ed. J. E. Birren and K. W. Schaie). Reinhold and van Nostrand, New York.

Martin, J. J. (1969). Thalamic syndromes. In *Handbook of clinical neurology* (ed. P. J. Vinken and G. W. Bruyn), Vol. 2, pp. 469–96. North-Holland, Amsterdam.

Mathew, N. T. and Meyer, J. S. (1974). Pathogenesis and natural history of transient global amnesia. *Stroke*, **5**, 303–11.

Maule, M. M., Milne, J. S., and Williamson, J. (1984). Mental illness and physical health in older people. *Age and Ageing*, **13**, 349–56.

Mayer-Gross, W., Slater, E., and Roth, M. (1954). *Clinical psychiatry*. Cassell and Company, London.

Mayfield, D., McLeod, G., and Hall, P. (1974). The CAGE questionnaire: validation of a new alcohol screening questionnaire. *American Journal of Psychiatry*, **131**, 1121–3.

McCartney, J. R. and Palmateer, L. M. (1985). Assessment of cognitive deficits in geriatric patients: a study of physician behavior. *Journal of the American Geriatrics Society*, **33**, 467–71.

McGinty, D. and Szymusiak, R. (1989). The basal forebrain and slow wave sleep: Mechanistic and functional aspects. In *Slow wave sleep: physiological, pathophysiological and functional aspects* (ed. A. Wauquier, C. Dugovic, and M. Radulovaci) pp. 61–73. Raven Press, New York.

Medina, J. L., Rubino, F. A., and Ross, E. (1974). Agitated delirium caused by infarctions of the hippocampal formation and fusiform and lingual gyri: a case report. *Neurology*, **24**, 1181–3.

Medina, J. L., Chokroverty, S. and Rubins, F. A. (1977). Syndrome of agitated delirium and visual impairment: a manifestation of medial temporo-occipital infarction. *Journal of Neurology, Neurosurgery and Psychiatry*, **40**, 861–4.

Menza, M. A., Murray, G. B., Holmes, V. F., and Rafuls, W. A. (1988). Controlled study of extrapyramidal reactions in the management of delirious, medically ill patients: intravenous haloperidol versus intravenous haloperidol plus benzodiazepines. *Heart Lung*, **17**, 238–41.

Mesulam, M. M., Waxman, S. G., Geschwind, N. *et al.* (1976). Acute confusional states with right middle cerebral artery infarctions. *Journal of Neurology, Neurosurgery and Psychiatry*, **39**, 84–9.

Mesulam, M. M. (1981). A cortical network for directed attention and unilateral neglect. *Annals of Neurology*, **10**, 309–25.

Mesulam, M. M. (1985). Attention, confusional states and neglect. In *Principles of behavioural neurology* (ed. M. M. Mesulam) pp. 125–68. F. A. Davis, Philadelphia.

Milam, S. B. and Bennett, C. R. (1987). Physostigmine reversal of drug-induced paradoxical excitement. *International Journal of Oral and Maxillofacial Surgery*, **16**, 190–3.

Millar, H. R. (1981). Psychiatric morbidity in elderly surgical patients. *British Journal of Psychiatry*, **138**, 17–20.

Miller, P. S., Richardson, J. S., Jyu, C. A., Lemay, J. S., Hiscock, M., and Keegan, D. L. (1988). Association of low serum anticholinergic levels and cognitive impairment in elderly presurgical patients. *American Journal of Psychiatry*, **145**, 342–5.

Modai, I., Beigal, Y., and Cygielman, G. (1986). Urinary amine metabolite excretion in a patient with adrenergic hyperactivity state: Reaction to phenelzine withdrawal and combined treatment. *Journal of Clinical Psychiatry*, **47**, 92–3.

Morgan, K. (1987). *Sleep and ageing*. Croom Helm, London.

Mori, E. and Yamadori, A. (1987). Acute confusional state and acute agitated delirium. Occurrence after infarction in the right middle cerebral artery territory. *Archives of Neurology*, **44**, 1139–43

Moroni, F., Cheney, D. L., and Costa, E. (1978). The turnover rate of acetylcholine in brain nuclei of rats injected intraventricularly and intraseptally with alpha – and beta-endorphin. *Neuropharmacology*, **17**, 191–6.

Morse, R. M. and Litin, E. M. (1969). Post-operative delirium: a study of etiologic factors. *American Journal of Psychiatry*, **126**, 388–95.

Muller, H. F. and Schwartz, G. (1978). Electroencephalograms and autopsy findings in geropsychiatry. *Journal of Gerontology*, **33**, 504–13.

Neal, M. J. (1976). Acetylcholine as a retinal transmitter substance. In *Transmitters in the visual process* (ed. S. L. Bonting) pp. 127–143. Pergamon Press, Oxford.

Obrecht, R., Okhomina, F. O. A., and Scott, D. F. (1979). Value of EEG in acute confusional states. *Journal of Neurology, Neurosurgery and Psychiatry*. **42**, 75–7.

Obrist, W. D., Sokoloff, L., Larson, N. A. *et al.* (1963). Relation of EEG to cerebral blood flow and metabolism in old age. *Electroencephalography and Clinical Neurology*, **15**, 610–19.

Obrist, W. D. (1979). Electroencephalographic changes in normal aging and dementia. In *Brain function in old age* (ed. F. Hoffmeister and C. Muller). Springer-Verlag, Berlin.

Olson, P. R., Suddeth, J. A. and Peterson, P. J. (1985). Hallucinations of widowhood. *Journal of the American Geriatrics Society*, **33**, 543–7.

Parkes, C. M. (1985). Bereavement. *British Journal of Psychiatry*, **146**, 11–17.

Patterson, A. and Zangwill, O. L. (1944). Recovery of spatial orientation in the post-traumatic confusional state. *Brain*, **67**, 54–68.

Pattie, A. H. and Gilleard, C. J. (1979). *Manual of the Clifton Assessment Procedures for the Elderly (CAPE)*. Hodder and Stoughton, Bucks.

Petrie, W. M. and Ban, T. A. (1981). Propanolol in organic agitation. *Lancet*, **i**, 324.

Pfeiffer, E. (1975). A short portable mental status questionnaire for the assessment of organic brain deficit in elderly patients. *Journal of the American Geriatrics Society*, **23**, 433–41.

Preskorn, S. H. and Denner, L. J. (1977). Benzodiazepines and withdrawal psychosis. Report of three cases. *Journal of the American Medical Association*, **237**, 36–8.

Pro, J. D. and Wells C. E. (1977). The use of the electroencephalogram in the diagnosis of delirium. *Diseases of the Nervous System*, **38**, 804–8.

Rabins, P. V. and Folstein, M. F. (1982). Delirium and dementia: Diagnostic criteria and fatality rates. *British Journal of Psychiatry*, **140**, 149–53.

Rau, R. and Voegt, H. (1972). Listeria meningoencephalitis and septicaemia in advanced age. *Medizinische Welt* (Stuttgart), **23**, 613–6.

Rees, W. D. (1971). The hallucinations of widowhood. *British Medical Journal*, **iv**, 37–41.

Regier, D. A., Boyd, J. H., Burke, J. D. *et al.* (1988). One-month prevalence of mental disorders in the United States. *Archives of General Psychiatry*, **45**, 977–86.

Reisberg, B., Ferris, S. H., Delean, M. D., and Crook, T. (1982). The global deterioration scale for assessment of primary degenerative dementia. *American Journal of Psychiatry*, **139**, 1136–9.

Richardson, J. T. E., Frith, C. D., Scott, E., *et al.* (1985). The effects of intravenous diazepam and hyoscine upon recognition memory. *Behavioral Brain Research*, **14**, 193–9.

Rigby, J., Harvey, M., and Davies, D. R. (1989). Mania precipitated by benzodiazepine withdrawal. *Acta Psychiatrica Scandinavica*, **79**, 406–7.

Roberts, A. H. (1963). The value of ECT in delirium. *British Journal of Psychiatry*, **109**, 653–5.

Robinson, G. W. (1939). Acute confusional states of old age. *Southern Medical Journal*, **32**, 479–86.

Robinson, G. W. (1956). The toxic delirious reactions of old age. In *Mental disorder in later life* (ed. D. J. Kaplan). Stanford University Press, Stanford.

Robinson, S. E. (1985). Cholinergic pathways in the brain. In *Central cholinergic mechanisms and adaptive dysfunctions* (ed. M. M. Singh, D. M. Warburton, and H. Lal) pp. 37–61. Plenum Press, New York.

Rockwood, K. (1989). Acute confusion in elderly medical patients. *Journal of the American Geriatrics Society*, **37**, 150–4.

Romano, J. and Engel, G. L. (1944). Delirium: I – EEG data. *Archives of Neurology and Psychiatry*, **51**, 356–92.

Ron, M. A. (1986). Volatile substance abuse: a review of possible long-term neurological, intellectual and psychiatric sequelae. *British Journal of Psychiatry*, **148**, 235–46.

Rose, P. E., Johnston, S. A., Meakin, M., Mackie, P. H., and Stuart, J. (1981). Serial study of C-reactive protein during infection in leukaemia. *Journal of Clinical Pathology*, **34**, 263–6.

Rosenbloom, A. (1988). Emerging treatment options in the alcohol withdrawal syndrome. *Journal of Clinical Psychiatry*, **49**, Supp. 28–32.

Roth, M. (1955). Natural history of mental disorder in old age. *Journal of Mental Science*, **101**, 281–301.

Roth, M. and Rosie, J. M. (1953). The use of ECT in mental disease with clouding of consciousness. *Journal of Mental Science*, **99**, 103.

Roth, M., Huppert, F. A., Tym, E. and Mountjoy, C. Q. (1988). *CAMDEX: The Cambridge Examination for Mental Disorders of the Elderly*. Cambridge University Press, Cambridge.

Rovner, B. W., Kafonek, S., Filipp, L., Lucas, M. J. and Folstein, M. F. (1986). Prevalence of mental illness in a Community Nursing Home. *American Journal of Psychiatry*, **143**, 1446–9.

Royal College of Physicians (1981). Organic mental impairment in the elderly. *Journal of the Royal College of Physicians*, **15**, 141–67.

Sakai, K. (1980). Some anatomical and physiological properties of ponto-mesen-cephalic tegmental neurons with special reference to the PGO waves and postural atonia during paradoxical sleep in the cat. In *The reticular formation revisited* (ed. J. A. Hobson and M. A. Brazier) pp. 427–47. Raven Press, New York.

Salzman, C. (1987). Treatment of the elderly agitated patient. *Journal of Clinical Psychiatry*, **48** Supp., 19–22.

Santamaria, J., Blesa, R., and Tolosa, E. S. (1984). Confusional syndrome in thalamic stroke. *Neurology*, **34**, 1618.

Saunders, P. A., Copeland, J. R. M., Dewey, M. E. *et al.* (1989). Alcohol use and abuse in the elderly: Findings from the Liverpool Longitudinal Study of continuing health in the community. *International Journal of Geriatric Psychiatry*, **4**, 103–8.

Schafer, D. F. and Jones, E. A. (1982). Hepatic encephalography and the gamma-aminobutyric acid neurotransmitter system. *Lancet*, **i**, 18–20.

Schmidt, D. (1982). *Adverse effects of anti-epileptic drugs*. Raven, New York.

Scott, J. (1960). Postoperative psychosis in the aged. *American Journal of Surgery*, **10**, 38–42.

Selzer, M. L. (1971). The Michigan Alcoholism Screening Tests: the quest for a new diagnostic instrument. *American Journal of Psychiatry*, **127**, 1653–1658.

Sendbuehler, J. M. and Goldstein, S. (1977). Attempted suicide among the aged. *Journal of the American Geriatrics Society*, **25**, 245–8.

Service, F. J., Dale, A. J., Elveback, L. R. *et al.* (1976). Insulinoma: clinical and diagnostic features of 60 consecutive cases. *Mayo Clinical Proceedings*, **51**, 417–29.

Seymour, D. G., Henschke, P. J., Cape, R. D. T. *et al.* (1980). Acute confusional states and dementia in the elderly : the role of dehydration/volume depletion, physical illness and age. *Age and Ageing*, **9**, 137–46.

Seymour, D. G. and Pringle, R. (1983). Post-operative complications in the elderly surgical patient. *Gerontology*, **29**, 262–70.

Seymour, D. G. and Vaz, F. G. (1987). Aspects of surgery in the elderly: pre-operative medical assessment. *British Journal of Hospital Medicine*, **37**, 102–8.

Shaw, P. J., Bates, D., Cartilidge, N. E. F. *et al.* (1986). Early intellectual dysfunction following coronary bypass surgery. *Quarterly Journal of Medicine*, **255**, 59–68.

Sheref, S. E. (1985). Pattern of CNS recovery following reversal of neuromuscular blockade. Comparison of atropine and glycopyrrolate. *British Journal of Anaesthesia*, **57**, 188–91.

Sherlock, S. P. V. (1985). *Diseases of the Liver and Biliary System*, (7th edn). Blackwell, Oxford.

Shulman, K. I. (1986). Mania in old age. In *Affective disorders in the elderly* (ed. E. Murphy) pp. 203–16. Churchill Livingstone, London.

Shulman, K. I., Shedletsky, R., and Silver, I. L. (1986). The challenge of time: clock-drawing and cognitive function in the elderly. *International Journal of Geriatric Psychiatry*, **1**, 135–40.

Siesjo, B. K., Johannsson, H., Norberg, K., and Salford, L. (1976). Brain function, metabolism and blood flow in moderate and severe arterial hypoxia. In *Brain work: proceedings of the Alfred Benzon Symposium VIII* (ed. D. H. Ingvar and N. A. Lassen) pp. 101–19. Munksgaard, Copenhagen.

Simon, A. and Cahan, R. B. (1963). The acute brain syndrome in geriatric patients. *Psychiatric Research Reports*, **16**, 8–21.

Simpson, C. J., and Kellett, J. M. (1987). The relationship between pre-operative anxiety and post-operative delirium. *Journal of Psychosomatic Research*, **31**, 491–7.

Singh, M. M. and Kay, S. R. (1985). Pharmacology of central cholinergic mechanisms and schizophrenic disorders. In *Central cholinergic mechanisms and adaptive dysfunctions* (ed. M. M. Singh, D. M. Warburton, and H. Lal) pp. 247–308. Plenum Press, New York.

Singh, M. M. (1985). Cholinergic mechanisms, adaptive brain processes and psychopathology: Commentary and a blueprint for research. In *Central cholinergic mechanisms and adaptive dysfunctions* (ed. M. M. Singh, D. M. Warburton, and H. Lal, H) pp. 353–97. Plenum Press, New York.

Sirois, F. (1988). Delirium: 100 cases. *Canadian Journal of Psychiatry*, **33**, 375–8.

Skegg, D. C. G., Doll, R., and Perry, J. (1977). Use of medicines in general practice. *British Medical Journal*, **i**, 1561–3.

Smith, J. S. and Brandon, S. (1970). Acute carbon monoxide poisoning: 3 years experience in a defined population. *Postgraduate Medical Journal*, **46**, 65–70.

Smith, L. W. and Dimsdale, J. E. (1989). Postcardiotomy delirium: Conclusions after 25 years? *American Journal of Psychiatry*, **146**, 452–8.

Spehlmann, R. (1981). *EEG primer*. Elsevier/North-Holland, Amsterdam.

Spivak, J. L. and Jackson, D. L. (1977). Pellagra: an analysis of 18 patients and a review of the literature. *Johns Hopkins Medical Journal*, **140**, 295–309.

Steinhart, M. J. (1979). Treatment of delirium – a reappraisal. *International Journal of Psychiatric Medicine*, **9**, 191–7.

Steriade, M. (1983). Cellular mechanisms of wakefulness and slow wave sleep. In *Sleep mechanisms and functions* (ed. A. Mayes) pp. 161–216. Van Nostrand Reinhold, Wokingham.

Stevenson, I. H., Salen, S. A. M. and Sheperd, A. M. M. (1979). Studies on drug absorption and metabolism in the elderly. In *Drugs and the elderly – perspectives in geriatric chemical pharmacology*. (ed. J. Crooks and I. H. Stevenson) pp. 51–63. Macmillan, London.

Strumpf, N. E. and Evans L. K. (1988). Physical restraint of hospitalised elderly: perceptions of patients and nurses. *Nursing Research*, **37**, 132–7.

Summers, W. K. and Reich, T. C. (1979). Delirium after cataract surgery: review and two cases. *American Journal of Psychiatry*, **136**, 386–91.

Sunderland, T., Tariot, P. N., Cohen R. M. *et al.* (1987). Anticholinergic sensitivity in patients with DAT and age-matched controls: A dose-response study. *Archives of General Psychiatry*, **44**, 418–26.

Swartz, M. S., Henschen, G. M., Cavenar, J. O. *et al.* (1982). A case of intermittent delirious mania. *American Journal of Psychiatry*, **139**, 1357–8.

Swigar, M. E., Benes, F. M., Rothman, S. L. G. *et al.* (1985). Behavioural correlates of computerised tomographic (CT) scan changes in older psychiatric patients. *Journal of the American Geriatrics Society*, **33**, 96–103.

Thomas, R. I., Cameron, D. J., and Fahs, M. C. (1988). A prospective study of delirium and prolonged hospital stay. *Archives of General Psychiatry*, **45**, 937–40.

Toro, G. and Román, G. (1978). Cerebral malaria. A dissociated vasculomyelinopathy. *Archives of Neurology*, **35**, 271–5.

Trzepacz, P. T., Baker, R. W., and Greenhouse, J. (1988). A symptom rating scale for delirium. *Psychiatry Research*, **23**, 89–97.

Tune, L. E., Holland, A., Folstein, M. F., Damlouji, N. F., Gardner, T. J., and Coyle, JT. (1981). Association of postoperative delirium with raised serum levels of anticholinergic drugs. *Lancet*, **ii**, 651–3.

Van Deuren, H. and Missotten, L. (1979). Atropine intoxication and the acute delirium of the elderly blind patient. *Bulletin dela Societé Belge d'Opthalmologie*, **186**, 27–9.

Vetter, N. J. and Ford, D. (1989). Anxiety and depression scores in elderly fallers. *International Journal of Geriatric Psychiatry*, **4**, 159–64.

Victor, M., Adams, R. D. and Collins, G. H. (1971). *The Wernicke-Korsakoff Syndrome*. Blackwell Scientific Publications, Oxford.

Von Cramon, D. Y., Hebel, N., and Schuri, U. (1985). A contribution to the anatomical basis of thalamic amnesia. *Brain*, **108**, 993–1008.

Von Korff, M. R., Eaton, W. W., and Keyl, P. M. (1985). The epidemiology of panic attacks and panic disorder: results of three community studies. *American Journal of Epidemiology*, **122**, 970–81.

Von Sweden, B. and Mellerio, F. (1988). Toxic ictal confusion: A symptomatic, situation-related subtype of nonconvulsive "absence" status epilepticus. *Journal of Epilepsy*, **1**, 157–62.

Warburton, D. M. and Wesnes, K. (1985a). Acetyl choline and attentional disorder. In *Central cholinergic mechanisms and adaptive dysfunctions* (ed. M. M. Singh, D. M. Warburton, and H. Lal) pp. 223–45. Plenum Press, New York.

Warburton, D. M. and Wesnes, K. (1985b). Historical overview of research on cholinergic systems and behaviour. In *Central cholinergic mechanisms and adaptive dysfunctions* (ed. M. M. Singh, D. M. Warburton, and H. Lal) pp. 1–35. Plenum Press, New York.

Ward, N. G., Rowlett, D. B., and Burke, P. (1987). Sodium amylobarbitone in the differential diagnosis of confusion. *American Journal of Psychiatry*, **135**, 75–8.

Whitaker, J. J. (1989). Postoperative confusion in the elderly. *International Journal of Geriatric Psychiatry*, **4**, 321–6.

Whittle, J. L. and Bates, J. H. (1979). Thermoregulatory failure secondary to acute illness. *Archives of Internal Medicine*, **139**, 418–21.

Willanger, R. and Klee, A. (1966). Metamorphopsia and other visual disturbances with latency occuring in patients with diffuse cerebral lesions. *Acta Neurologica Scandinavica*, **42**, 1–18.

Willi, J. (1966). Delir, Dammerzustand, und Verwirrtheit bei körperlich Kranken. In *Akute Psychische Begleiterscheinungen Körperlicher Krankheiten.*(ed. M. Bleuler, J. Willi, and H. R. Buehler) pp. 27–158. Thieme Verlag, Stuttgart.

Wolff, H. G. and Curran, D. (1935). The nature of delirium and allied states: the dysergastic reaction. *Archives of Neurology and Psychiatry*, **33**, 1175–1215.

Wolfson, P., Cohen, M., Lindesay, J., and Murphy, E. (1990). Section 47 and its use among people with mental disorders. *Journal of Public Health Medicine*, **12**, 9–14.

Woodhead, M. A. and Macfarlane, G. T. (1987). Comparative clinical and laboratory features of Legionella with pneumococcal and mycoplasma pneumonias. *British Journal of Diseases of the Chest*, **81**, 133–9.

World Health Organization (1972). *Psychogeriatrics – Technical report 507*. World Health Organization, Geneva.

World Health Organization (1977). *Manual of the International Statistical Classification of Diseases, Injuries, and Causes of Death* (9th revision). World Health Organization, Geneva.

Wragg, R. E., Dimsdale, J. E., Moser, K. M., *et al.* (1988). Operative predictors of delirium after pulmonary thromboendarterectomy. A model for postcardiotomy delirium? *Journal of Thoracic Cardiovascular Surgery*, **96**, 524–9.

Zeneroli, M. L., Pinelli, G., Gollini, G. *et al.* (1984). Visual evoked potential: A diagnostic tool for the assessment of hepatic encephalopathy. *Gut*, **25**, 291–9.

Ziskind, E. (1965). An explanation of mental symptoms found in acute sensory deprivation:researches 1958–1963. *American Journal of Psychiatry*, **121**, 939–45.

Index

MAPPERLEY HOSPITAL MEDICAL LIBRARY

Author................ ....Class Mark................ ....

Title................ ....